Early Contractor Involvement

emerald
PUBLISHING

ice
Publishing

Early Contractor Involvement

Improving the management of contract risk

**Michael Smith, Matthew Finn,
Dr Jon Broome and Catherine Maddox**

Published by Emerald Publishing Limited, Floor 5,
Northspring, 21–23 Wellington Street, Leeds LS1 4DL.

ICE Publishing is an imprint of Emerald Publishing Limited

Other ICE Publishing titles:
Procurement and Contract Strategies for Construction
Ian Heaphy. ISBN 978-0-7277-6371-6
FIDIC 2017: The Contract Manager's Handbook
Geoffrey Smith. ISBN 978-0-7277-6652-6
*Building Regulations, Codes and Standards: A guide for safe,
sustainable and healthy development*
Mark Key. ISBN 978-0-7277-6535-2

A catalogue record for this book is available from the British Library

ISBN 978-1-83549-897-2

Cover photo: Niels Melander/Alamy Stock Photo

Commissioning Editor: Michael Fenton
Content Development Editor: Cathy Sellars
Books Production Lead: Emma Sudderick

Typeset by KnowledgeWorks Global Ltd.
Index created by David Gaskell

Contents

About the authors

Dr Jon Broome BEng PhD FAPM. Jon's expertise in contract strategy started to grow in 1998 when, having completed his PhD on the practical application of the NEC form of contract, he did two years of research in collaborative contracting strategies, resulting in his 2002 book *Procurement Routes for Partnering: a practical guide* (ICE Publishing). For the last 25+ years, Jon has been an independent consultant at Leading Edge Project Consulting Ltd, UK. While his 'home' sector is heavy engineering and construction, he has helped set up innovative commercial/contractual arrangements for projects in the ship-building, aerospace, defence and coal-mining sectors. In the heavy engineering and construction sector, Jon has helped set up numerous early contractor involvement and alliance arrangements. His involvement has ranged from helping develop the 'big-picture' contract strategy; to drafting contracts; refining technical documents and assisting in their set-up, as a trainer, facilitator and chair of regular meetings.

Jon is also well-known for his expertise in NEC contracts, a former long-term chair of the Association for Project Management's (APM) Contracts & Procurement Specific Interest Group and a former deputy chair of the APM itself.

Matthew Finn BSc Hons LLM FCIArb MCInstCES FCIOB FRICS MAE MEWI is a Senior Managing Director at Ankura, based in London, and leads the firm's quantum practice in construction, disputes and advisory. He is regularly appointed as an expert witness in the field of quantum (damages) in energy, infrastructure and construction matters. He has given oral evidence on many occasions in UK litigation and in many high-value international arbitrations (up to US$10 bn disputes) under both traditional cross-examination and under concurrent evidence by leading King's Counsel. He has submitted more than 100 quantum expert reports in litigation and in international arbitration under ICC, SIAC, UNCITRAL, LCIA, SCC and PCA rules. Matthew has been recognised in Who's Who Legal since 2018 and is ranked as a Global Elite Thought Leader in 2023 for arbitration, quantum and delay, and construction. He was the sole winner of the Arbitration Expert Witnesses UK at Lexology's Client Choice Awards 2021.

Matthew has worked in the construction industry as a chartered quantity surveyor and chartered construction manager in both consulting and contracting organisations. He has worked on a range of projects for major contractors and consultancies on projects for private, corporate and public body clients. He has worked on projects in the building, civil engineering, nuclear, rail, oil and gas, and building services sectors. In addition to expert witness appointments, Matthew is a certified civil and commercial mediator, construction adjudicator, international and domestic arbitrator.

Catherine Maddox is a Senior Associate at Ashurst LLP in London. Catherine specialises in public procurement law and qualified as a solicitor (England and Wales) in 2015. She advises central and local government authorities, other public sector bodies and utilities on the application of public procurement rules. Catherine advises on the structuring of complex procurement processes, including choice of tendering procedure, preparation of procurement notices, the drafting of tender documentation and obligations during the award and debriefing stages.

She is currently advising Great British Nuclear on the procurement of the major works packages for its forthcoming small modular nuclear reactor programme.

Michael Smith is a Senior Consultant at Ashurst LLP in London. He has over 30 years of experience, advising on the construction aspects of major infrastructure project procurement, domestic and international EPC contracting and all other areas of noncontentious construction and engineering law, with a particular focus on the thermal, nuclear and renewable energy sectors, including energy from waste, solar and on- and offshore wind farm development.

His clients include the UK Government, funders, sponsors and contractors. Michael has advised on nationally significant infrastructure projects including the Hinkley Point C nuclear power project, the HS2 rail project, the Thames Tideway Tunnel project, the East West Rail project, the Edinburgh Tram project and the Moray Firth East and West offshore wind farm projects. He is currently advising Great British Nuclear on the procurement of the major works packages for its forthcoming small modular nuclear reactor programme.

emerald
PUBLISHING

ice
Publishing

Michael Smith, Matthew Finn and Jon Broome
ISBN 978-1-83549-897-2
https://doi.org/10.1108/978-1-83549-894-120241001

Chapter 1
Construction procurement models: options and trends

Michael Smith

Abstract

This chapter describes
- the various construction procurement options currently in widespread use within the UK
- the factors that should inform a decision on the most appropriate construction procurement option for any given project
- current market procurement trends and the reasons why a two-stage contracting model should now gain more traction within the UK infrastructure sector than it has done to date.

1.1. Introduction

Previous experience within the UK construction industry has led all project stakeholders, including the UK Government, to question whether the previously widespread use of a single, lump-sum, turnkey contract to procure the development of infrastructure assets, often by way of a limited recourse project financing structure, is in the best, long-term interest of all project participants.

Until fairly recently, those funding or sponsoring such projects tended to support the view that the use of a single, turnkey, design and construction contract was the only appropriate procurement choice, unless they were unable to find a contractor willing to wrap that degree of risk transfer on a particular project. Looked at from a purely legal perspective and while standing in the shoes of a risk-adverse project sponsor or lender, it was difficult to counsel against that conclusion.

However, just as the rest of the world is beginning to embrace the limited recourse project financing models developed in the UK over the last 20 years, we are changing direction. While it is not unusual for the UK to head in a different direction from the rest of Europe, on this occasion there may be consensus that it is doing so for the right reasons.

The UK Government has always been concerned about the adversarial nature of the UK construction industry. Ever since Sir Michael Latham focused attention on it in his ground-breaking report (*Constructing the Team*, Latham, 1994), there has been a concern that the industry uses contracting arrangements that are adversarial in nature. As such, they promote disputes and do not operate in the best interests of both parties. That concern has been significantly addressed in the intervening years, not least by the advent of the NEC suite of standard form contracts.

However, other developments within the industry, including the Government's sponsorship of the private finance initiative (PFI) to procure public infrastructure, have given rise to new concerns. The Government's support for public–private partnerships (PPPs) was eventually undermined by criticism from all sides that the model does not deliver value for money and concern at what were perceived to be excessive profits being made by the private sector. At the same time, some significant events (including the demise of Carillion and the difficulties which led to Interserve's exit from the waste to energy sector) demonstrated the potentially disastrous consequences of a single entity wrapping the entire project delivery risk without perhaps properly understanding the nature of that risk and making appropriate allowance for it. This led the Government to declare, at the end of 2018, that it would no longer use the PFI/PPP delivery model for the procurement of public infrastructure works.

In the report on its investigation into the rescue of Carillion's PFI hospital contracts, published in January 2020, the National Audit Office observed

> On the one hand fixed-price contracts, like PFIs, can ensure that contractors are both incentivised to manage costs and bear the brunt of cost rises. On the other hand, some risk ultimately remains with the public sector; when things go wrong beyond a certain tipping point, the public sector will bear the consequences. For these projects the tipping point was reached through a combination of high cost overruns on the construction and the failure of the contractor who had been expected to pay for them (National Audit Office, 2020: p. 6).

The National Audit Office stated that the cost over-runs were largely attributable to design problems at hospitals and observed

> The PFI companies and investors are meant to rely on their contractors to manage the design of the projects. The investors' payments to the PFI companies are made in line with technical adviser reports that track progress against plan and provide no assurance over the design. We have not looked at what went wrong with the design process within Carillion and its subcontractors. NHS England and NHS Improvement and a few individuals involved in the construction projects told us that one of the causes may have been that Carillion's original pricing was too low to meet the required specification (National Audit Office, 2020: p. 21).

As a consequence of these concerns, there has been a developing consensus within the UK that the Client side of the construction industry needs to engage earlier and better with the whole of its supply chain in order to properly understand and manage the particular construction risks on any given project. Stepping down the whole of that risk to a single contractor who has priced that risk in a highly competitive environment may no longer be seen as representing value for money and may even be considered unwise in certain technologically challenging sectors.

The authors of this publication believe that this consensus is now leading to the more widespread use of a two-stage contract delivery model. This model is most commonly based on the use of a standard form contract – the NEC4 Engineering and Construction Contract (NEC, 2017), incorporating the main pricing Option C (Target Contract with Activity Schedule) and the Secondary Option X22 (Early Contractor Involvement). The Contractor is reimbursed on a 'cost plus (fee)' basis. A two-stage appointment process, which provides for early contractor involvement (ECI) during the stage one design phase and a target/outturn construction cost incentive scheme during the stage two construction phase, is then used in order to

- interrogate, understand and manage out construction risk before the parties commit to a target price for the stage two construction phase
- manage the cost escalation risk that is inherent with such a delivery model.

This procurement model has been used and refined by those involved in some of the largest civil works infrastructure projects within the UK over the last ten years – projects on which, given the nature of the design and construction risk, there was little or no market appetite for a single, turnkey wrap.[1] This model and variants of it, including that referred to in the following paragraph, are explained in more detail in Chapter 4 *Managing cost escalation on a cost reimbursable contract.*

The use of such two-stage contract delivery models has been promoted by the UK Government for several years. As long ago as July 2014 the UK Government Cabinet Office produced guidance on three new models of construction procurement for public sector clients, including a 'two-stage open-book' model. Peter Hansford, the Government Chief Construction Adviser at the time, stated in the foreword to the guidance that the aim of the Government Construction Task Group was to

provide cost certainty, which is an essential element of providing better long-term value from the delivery of construction projects. It is vital that *clients* enter the procurement process knowing what their projects should cost and that the procurement vehicle adopted provides them with confidence of what their projects will cost (Cabinet Office, 2014: p. 1).

The guidance cited NEC3 Option C, the JCT Constructing Excellence Contract and the PPC 2000 (published by the Association of Consultant Architects (ACA)) as appropriate standard form contracts for the implementation of the proposed two-stage open-book model. Since then, the NEC has issued its Secondary Option X22 (Early Contractor Involvement) for use with the NEC4 Engineering and Construction Contract and the JCT has issued its Pre-Construction Services Agreement to serve as the first stage in a two-stage process.

1.2. Historical construction procurement options for major infrastructure projects

Set out below is a summary of the three forms of construction procurement that have been most commonly used across the various UK construction industry sectors to date. As we discuss in more detail below, the procurement options that are most commonly used within a particular industry sector will depend on the technical and commercial considerations relevant to that sector and the approach to risk allocation and the pricing thereof that is adopted by the project stakeholders who are most active within that sector.

1.2.1 Construct only

With this traditional form of procurement most, if not all, of the design is carried out by a design consultancy team engaged by the Client. The Contractor is then engaged to build to that design, although it may have responsibility for some design, sometimes known as 'the contractor's design portion'.

Examples of current standard form contracts which are most commonly used in the UK for this type of procurement include
- FIDIC 2017 Red Book (FIDIC, 2017)
- NEC4 Engineering and Construction Contract, Option A or B (Priced Contract) (NEC, 2017)
- JCT Standard Building Contract, 2016 Edition (JCT, 2016).

1.2.2 Design and build

The main contractor is responsible for the design and construction of the works, thereby creating a single point of responsibility for both the design and construction of the whole of the works – sometimes referred to as a 'turnkey' contract or (in the process engineering sector, in particular) an 'EPC' contract – that is, an engineering, procurement and construction contract.

As the Contractor will be responsible for the whole of the design of the works under this procurement model, if any of the output design requirements included within the works specification were actually prepared by a design consultant appointed by the Client, the Contractor may insist on the novation of that consultant's appointment agreement so that it has a contractual nexus with the author of that design.

Variant forms of this procurement approach, whereby the Client retains responsibility for the output design requirements included within the works specification, with the Contractor responsible for the detailed design development and construction of the works, are usually referred to as 'design and build' or 'design and construct' contracts, in order to distinguish them from single point, 'turnkey' contracts.

Examples of current standard form contracts which are most commonly used in the UK for this type of procurement include
- FIDIC 2017 Silver Book, Turnkey Contract
- FICIC 2017 Yellow Book, Design and Build Contract
- NEC4 Engineering and Construction Contract, Option A or B (Priced Contract) or Option C (Target Contract)
- JCT Major Project Contract, 2016 Edition
- JCT Design and Build Contract, 2016 Edition.

1.2.3 Construction management

The Client appoints a number of different contractors and/or consultants to carry out the various elements of the works, which are sometimes referred to as 'packages', as well as a construction manager whose primary role is to manage the construction interface between the various package contractors on behalf of the Client. Package contracts may include design consultancy services or construction works or may be let on a design and build basis.

Examples of current standard form contracts which are most commonly used in the UK for this type of procurement include
- FIDIC 2017 Yellow Book, Design and Build Contract for package contracts and FIDIC 2017 White Book, Model Services Agreement for the construction manager's appointment
- NEC4 Engineering and Construction Contract, Option A or B (Priced Contract), Option C or D (Target Contract) or Option E (Cost Reimbursable Contract) for package contracts and NEC4 Professional Services Contract for the construction manager's appointment
- JCT Construction Management Trade Contract and Construction Management Appointment, 2016 Editions.

1.3. Key factors that will influence the choice of construction procurement option

1.3.1 Time, cost and quality

Historically, most professionals operating within the construction industry have been taught that procurement decisions should largely be based on the need to balance the three (mostly competing) points of the 'procurement triangle', comprising time, cost and quality. As a theoretical consideration, that is undoubtably as true now as it ever was.

By way of example, Figure 1.1 seeks to summarise, in very simple terms, some of the theoretical advantages and disadvantages of a single, lump sum, turnkey contract on the one hand and a multicontract construction management procurement on the other hand.

Figure 1.1 Advantages and disadvantages of a turnkey contract against a multicontract construction management procurement (author's own)

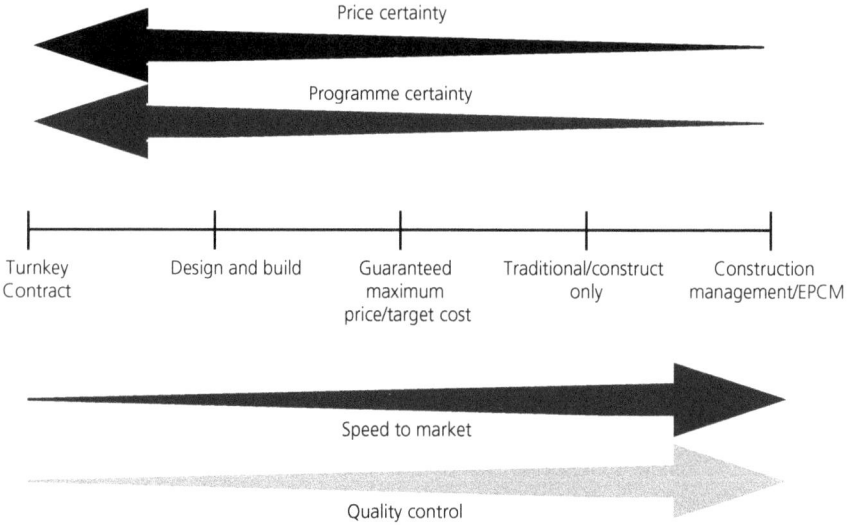

In practice, however, the Client's decision on how best to balance its competing time, cost and quality requirements is likely to be driven by the key considerations, shown in Tables 1.1 and 1.2.

Table 1.1 Key considerations for single contract turnkey models (continued on next page)

Advantages	Disadvantages
Single point responsibility for design and construction.	The Contractor largely controls the design development process and therefore the detailed design solution.
The Client's risk is limited: the Contractor takes on the majority of the risk and does so for a fixed price.	The Contractor largely controls the selection and appointment of its supply chain.

Table 1.1 Continued

Advantages	Disadvantages
Bankability: a fixed, lump sum price and fixed date(s) for completion of the works facilitate the financial modelling on which the financing of the project is based.	Assumption of risk attracts risk premia forcing up contract price in a manner which may not represent value for money for the Client.
Performance of the completed works/asset and protection of the Client's revenue stream: subject to appropriate liability limitation provisions, the Contractor will usually guarantee the performance and/or availability of the facility which comprises the completed works and pay liquidated damages in respect of any shortfall in the guaranteed/modelled performance and/or availability.	All the Client's eggs are in one basket in the event of the Contractor's demise (although this risk is usually offset to a material degree by the Client affording itself the right to step into key subcontracts and by the provision of performance security in relation to the Contractor's liabilities on any consequent termination of the contract).

Table 1.2 Key considerations for multicontract construction management models

Advantages	Disadvantages
The Client has greater control over works packaging and design development processes. It has the flexibility to choose its own supply chain, rather than be tied to that of a turnkey contractor (whose best interests may be served by its relationship with its subcontractors rather than the Client).	The Client has multiple points of accountability that need to be proactively and effectively managed. This may make it more difficult for the Client to secure third-party finance for its project (unless it is able to evidence that it has the experience, expertise and management resource necessary to mitigate and manage the elevated risk of multiple interfaces and disputes).
The Client can begin construction earlier than would otherwise be the case – design development and construction of any parts of the works which are not interdependent can be carried out concurrently.	The Client retains the interface management risk between the various works packages and is usually able to step down only a very limited amount of that risk to its construction manager. Including an effective incentive scheme across the various works packages and perhaps including the construction manager within that scheme may go some way to mitigating this risk, as discussed in Section 4.3 of Chapter 4.
Early involvement of supply chain facilitates value engineering, 'buildability' analysis, pricing discounts and avoidance of a turnkey contractor's mark-up.	Fixed, lump sum prices may be difficult to obtain across the package contracts.

1.3.2   Project finance

If the Client is seeking to raise debt finance to fund any part of the project capex on a limited recourse basis so that, at least initially (i.e. until the works are complete), there is no asset providing security to the funders for repayment of the debt, the funders are likely to insist on a procurement structure which delivers the greatest degree of certainty on project cost and programme.

Traditionally that has meant a turnkey or design and build contract solution, although there have been some notable recent exceptions to this general rule. The absence of contractors willing to provide a turnkey 'wrap' solution in the offshore wind sector has led to certain market-leading developers securing debt finance on what are essentially construction management procurements. Debt finance has also been raised for construction management procurements in the solar sector, when the interface risk has been viewed by the funders as low and therefore manageable. It is also not uncommon to see construction management procurements funded in the oil and gas sector, where the project sponsors are used to providing completion guarantees in order to secure that finance.

On a corporate finance project, the Client will have greater flexibility to choose the procurement structure which prioritises its commercial objectives. It is therefore likely to focus more on a structure which reduces the initial capex of the project, particularly if it has the available resources and expertise necessary to deliver a construction management or engineering, procurement and construction management (EPCM) procurement.

Looking forward, it appears increasingly likely that the UK Government will seek to use the regulated asset base (RAB) funding model to secure private sector finance for major, greenfield infrastructure projects in the nuclear energy and aviation sectors. The RAB model has been used elsewhere in the world to secure major infrastructure for regulated utilities but the UK Government is currently developing its own model in consultation with the likely project sponsors and investors. It is anticipated that, under its RAB model, those investing in the project will earn a return on their investment during the construction phase and those carrying out the work will be incentivised to control unnecessary costs escalation by the use of target price contracting.

1.3.3   Client resources

The extent of the management resource available to the Client, either internally or externally by way of a consultancy support services agreement, should be a key factor in the Client's decision making process. Anecdotal evidence from some of the earlier construction management procurements in the UK has led to the conclusion among most commentators that the success of this more labour-intensive procurement option is primarily dependent on the Client's level of resource and track record in that type of procurement.

1.3.4   Client risk appetite

The Client's approach to risk management will also be a key factor in determining how it wishes to structure the procurement. Normally the Client (and/or its funders) will prefer to clearly step down responsibility for the following key risks to whichever party is best placed to manage that risk

- adequacy of the output design requirements
- development of the detailed design requirements
- compliance with the design requirements and completion of the construction works
- management of the programme and operational interface between the various project participants
- costs escalation (save to the extent caused by Client default and certain specified perils)
- performance of the completed works/asset.

However, for many years the UK Government has been trying to persuade the private sector to adopt a more consensual approach to risk allocation and most of the standard form construction contracts now include collaborative contracting provisions in an effort to promote a partnering approach. Should the Client wish to adopt some of the more innovative alliancing arrangements that are addressed elsewhere in this chapter (pursuant to which it will treat its supply chain as a group of partners who work together with the Client's representative to manage, mitigate and share the key project risks), then it (and all other alliance participants) will need to secure the appropriate resources to manage that alliance effectively.

1.3.5 Supply chain risk appetite

Having come to an initial conclusion on the most appropriate procurement structure, the Client will then need to consider whether the prevailing market conditions will facilitate its delivery. In particular the Client will wish to consider and understand

- the extent to which there is sufficient supply chain appetite within the relevant sector to accept the risks being stepped down and to price those risks on a competitive basis
- whether the relevant contractor is able to demonstrate to the Client (and its funders) that it has the requisite expertise, technical resource and financial substance to properly manage the risk which it has agreed to accept.

So, when developing its tender procurement strategy, it will be important for the Client to understand the market from which its supply chain will be drawn. In Chapter 3 *Selection process* we consider in more detail how the Client might do that.

1.4. Deciding on the correct construction procurement option

1.4.1 Lump sum and/or cost reimbursable?

One of the first decisions to be made on any project procurement is the packaging and pricing strategy. In particular, if the whole of the design and construction is not to be 'wrapped' within a single, lump sum, fixed price, 'turnkey' contract, how many principal works package contracts will there be and which pricing option within each contract is most likely to deliver value for money for the Client? Should each contract be

▩ a fixed price, lump sum contract, under which the fixed price is only adjusted in respect of agreed compensation events or

▩ a cost reimbursable contract, under which the Contractor will be reimbursed those costs which it reasonably incurs in carrying out and completing the works, plus an additional fee to cover its overheads and its profit entitlement, calculated as a percentage of those costs?

1.4.2   Fixed price/lump sum contracts

If the Client's basic engineering design solution and technical requirements for the works can be sufficiently developed and interrogated prior to contract signature so that the Contractor can properly define and price the consequent work-scope without an undue risk margin allowance, then a lump sum, fixed price contract should deliver costs certainty for the Client and may well therefore represent value for money for the Client. In those circumstances either the FIDIC Silver Book or the NEC4 Engineering and Construction Contract, Option A (Priced Contract with Activity Schedule) are the most commonly used standard form contracts within the UK infrastructure sector.

If that is not the case then the Client may find that there is little or no contractor interest in the market to wrap turnkey risk within a fixed price contract or that the margin which the Contractor will charge in order to do so may give rise to value for money considerations. Even if there is sufficient market appetite to secure a competitive tender process, the Client will be well advised to ensure that its prequalification process adequately interrogates the ability of the tenderers to manage the risks (and potential liabilities) which the successful tenderer will assume. As mentioned earlier in this Chapter, the demise of Carillion and the difficulties which led Interserve to exit from the energy from waste sector emphasise the importance of doing so.

1.4.3   Cost reimbursable contracts

So, if

▩ programme constraints dictate that part of the Client's technical requirements for the works package can only be developed and interrogated post contract and/or

▩ there is little or no market appetite to accept (or sensibly price)

    o   the risk of an uncertain or technically challenging design and/or work-scope and/or

    o   the interface risk on a multi-contract construction management procurement,

then a cost reimbursable contract is likely to represent better value for money for the Client.

It may also be the case that a cost reimbursable contract is the only available procurement option for the Client, if there is no appetite to wrap the quantum of design and/or construction risk within the relevant market sector. Obvious recent examples of this are the huge civil works packages let on the Thames Tideway Tunnel, Hinkley Point C and HS2 projects.

If the Client does decide to adopt a cost reimbursable approach for its project, it will need to decide which standard form contract will give it the best chance to manage the risk of significant costs escalation to which it will then be exposed. Within the UK infrastructure sector, the current consensus is that the NEC4 Engineering and Construction Contract, Option C (Target Contract with Activity Schedule) will give it the best means of doing so. It contains all the project management tools that are most commonly used to manage the risk of unnecessary costs escalation. Some, but not all, of these project management tools have also now been adopted within the second editions of the FIDIC Red, Yellow and Silver Books (published in 2017). However, they do not provide for

- a two-stage ECI tender process, and
- a target price pain/gain share incentive regime
- both of which are included as standard options within the NEC4 Engineering and Construction Contract. These issues are discussed in more detail in Chapter 4 *Managing costs escalation on a cost reimbursable contract*.

1.5. Controlling costs escalation under a reimbursable contract

1.5.1 Facilitating proactive construction management

It is a widely held belief that effective costs management, particularly where the project is to be procured on a construction management basis (i.e. where the Client retains responsibility for managing the interface risk between several different works packages), will depend on the performance of a strong and effective construction management role on behalf of the Client. So, the Client will need to give itself (or its appointee) the project management tools which it needs in order to properly perform that role. The NEC4 conditions of contract do this by

- only allowing for the reimbursement of 'defined costs' and not 'disallowed costs' (thereby encouraging the Contractor to be efficient in the management of its supply chain)
- obliging the Contractor to comply with a key dates regime, an information release schedule and a site access protocol (thereby assisting the Client in its management of interface risk)
- promoting transparency between the parties through the use of a risk register, disclosure of the Contractor's pricing contingency allowance and agreement between the parties on float allowance and ownership
- facilitating good project management by way of the use of early warning notices, risk reduction meetings and programme updates.

While the second editions of the FIDIC Red, Yellow and Silver Books do now embrace some of these concepts, they are not as comprehensive in their coverage. The differences are discussed in more detail within Chapter 4 *Managing costs escalation on a cost reimbursable contract.*

1.5.2   Target price incentive schemes

In order to ensure that the Contractor is incentivised to work efficiently, in the best interests of the project, those tendering for a 'cost plus' contract will usually be asked to bid a target price for the works. The successful tenderer will then share an agreed percentage of any resultant 'pain' or 'gain' – that is, the amount by which the actual outturn construction cost is more or less than the target price bid by the Contractor.

Where this type of incentive scheme is adopted, most of the negotiations between the parties will usually concern

- the list of compensation events that will entitle the Contractor to an adjustment to the target price
- the percentages at which, and the bands within which, any pain or gain is shared between the parties
- the threshold circumstances in which the fee (i.e. the Contractor's overheads and profit) is no longer applied to any cost – that is, no longer payable to the Contractor
- the extent (if any) to which the Contractor's pain share counts towards any limit on its aggregate liability under the contract.

1.5.3   Two-stage contracts involving ECI

This model involves the parties entering into a single, two-stage, cost reimbursable contract pursuant to which the parties engage in an initial, collaborative process ('stage one') to develop the design, planning and programming for the project, before advancing to the second construction phase ('stage two').

A properly constructed, two-stage contract providing for ECI is likely to represent the parties' best opportunity to reduce their exposure to pain risk on a target price contract or even to achieve value for money in relation to a properly interrogated lump sum, turnkey contract price for stage two, on the basis that asking the Contractor to price an uncertain design solution under a single stage contract is unlikely to represent value for money.

Pursuant to the NEC's Secondary Option X22, which is designed for use with the NEC4 Main Option C Target Contract, the Client appoints the Contractor to undertake design development and ancillary services during stage one. The scope and duration of these services is defined in the Scope. Thereafter the Client will have the option of appointing the Contractor to finish the detailed design development and execute the works during stage two.

This process should enable the Client to

- manage out some of the potential pain share risk by investigating and developing the Contractor's proposed design solution during stage one
- retain some control over the process by which the design solution is developed and adopted for the project (thereby avoiding the danger of significant reputational damage that could occur if an inadequate design solution is adopted by a package contractor with no interest in the operational performance of the project)
- maximise the opportunities for value engineering and innovation, thereby minimising both capex and opex costs for the project by focusing the delivery team on the Client's objectives right from the outset
- integrate design development within construction planning at the earliest possible stage, thereby allowing more time for the parties to plan for critical events and prepare a fully detailed construction programme.

In Chapter 5 *Realising the practical benefits* we examine in more detail the particular benefits that should be delivered by an ECI process and the factors that will be key to the delivery of those benefits.

However, the Client should be mindful to safeguard against the risk of programme slippage and a gradual erosion of its bargaining power as stage one progresses. It can do this by

- including clear programme and work-scope requirements for stage one within the Invitation to Tender
- providing clearly for the Client's right to withdraw from the process without penalty at the end of stage one (and to proceed with the next best bid)
- requiring agreement on the stage two conditions of contract as a condition precedent to the stage one appointment
- maintaining competitive tension within the tender procedure by
  - o evaluating change to the stage two target price using the competitive pricing information submitted with the Contractor's original bid submission
  - o allowing the Contractor to share in any saving between the initial target price included within the Contractor's accepted tender and the sum of the revised target price fixed at the end of the stage one process and the amounts paid during stage one (albeit with payment of any such saving held over until completion of the works and the assessment of any pain share payable by the Contractor at that stage).

Amendment to the NEC4 Secondary Option Clause X22 will be required in order to reflect any of the above arrangements.

It should also be noted that the apparent intention of the NEC4 Secondary Option Clause X22 is to incentivise the Contractor to interface effectively with other third parties, by entitling the

Contractor to share in any saving between the outturn project costs and the budget for those project costs which was included in the Contract Data. If the NEC4 Secondary Option Clause X22 is to be used for a single contract procurement, the description of the budget in the Contract Data will therefore need to be amended so that it contains reference only to the budget cost for that contract and not for any other elements of the project.

1.5.4   FEED agreements

In the process engineering sector, front-end engineering and design (FEED) agreements are commonly used to regulate the development of front-end engineering and design proposals, sufficient to facilitate the obtaining of necessary project approvals by the Client and the establishment of a rough investment cost for the project. The FEED design process will also develop the technical requirements which will comprise the Client's requirements for the purposes of an EPC contract. As such they are also sometimes adapted to perform a similar role to the JCT Pre-Construction Services Agreement (which is more commonly used in the real estate sector) and the NEC4 Secondary Option Clause X22 (which is more commonly used in the infrastructure sector).

The perceived advantages of a properly constructed FEED process are therefore very similar. It should

- enable risks to be properly identified, assessed, priced and mitigated
- reduce the costs of tendering as only one design process is undertaken
- facilitate the achievement of value for money through early contractor involvement in design and pricing, on a transparent basis
- optimise construction efficiencies and reduce operating costs.

However, to deliver these benefits, a FEED process will usually need to recognise and embrace the following key principles

- it should facilitate outturn capex savings by an initial, up-front investment in value engineering
- it should recognise the need for relevant technical design expertise
- it should require and facilitate early involvement from the operation and maintenance contractor and other relevant stakeholders
- the services work-scope and deliverables should be tailored to the particular nature of the works package and/or project
- it should recognise the need to provide for whole life project and systems changes.

1.5.5   Alliancing arrangements on multicontract procurements

As discussed in more detail in Chapter 4 *Managing costs escalation on a cost reimbursable contract*, alliancing arrangements are now being used on some of the most significant civil engineering projects, including the Thames Tideway Tunnel and HS2 projects, to promote a collaboration incentive across different works packages on complex construction management procurements.

1.5.6   Partnering/collaboration arrangements

A partnering or collaboration arrangement is included within the NEC4 Engineering and Construction Contract, Secondary Option Clause X12 (Multiparty Collaboration). It is designed to promote collaboration across different works packages in order to facilitate the achievement of a common set of objectives set by the Client, which may include a target price for the project, a target date for project completion and/or the achievement of other key performance indicators.

The scheme is based on three fundamental principles
- a successful incentive regime depends on a successful alliance
- an alliance can only exist where there is consensus
- the alliance parties can only be incentivised to achieve an outcome if they are able to influence the achievement of that outcome by their performance.

Accordingly
- there is no Contractor exposure to any project target cost pain or any project programme over-run
- the Client has the right to add new incentives to the scheme or revise existing ones if their achievement can no longer be influenced
- the alliance is managed by consensus – that is, by a core group on which each partner is equally represented and decisions are made on a unanimous basis
- if any participant wishes to withdraw it may do so without penalty, provided it has acted in good faith.

Clearly, the effectiveness of such a collaboration scheme will depend on a continuing consensus between all the participants, driven by a collective gain share. It is of note, from a legal perspective at least, that the Client has very little power if the incentive for collaboration no longer exists and consensus breaks down. In such circumstances the Client will have no claim against the individual participants, unless they have acted in bad faith.

1.5.7   Alliance agreements

In July 2018 the NEC published the NEC4 Alliance Contract. The intention is that the Client and all key members of the supply chain enter into the agreement, under which all of them have an equal voice and share in the performance of the alliance as whole. Accordingly the Alliance Contract replaces each of the works package contracts, thereby removing any individual target price pain/gain share incentive schemes included within those works package contracts.

It would appear, if only anecdotally, that usage of the NEC4 Alliance Contract is not yet widespread although, according to the NEC

- Yorkshire Water Services adopted it for the creation of the Yorkshire Alliance, which will deliver part of its asset management period 7 programme
- Highways England and its six partners adopted it in April 2020 to develop, design and construct England's smart motorways programme
- Sheffield Hallam University used it for the first phase of a £700 m, 20-year redevelopment of its city centre campus
- EDF and its Chinese partner CGN adopted it in September 2018 to deliver an estimated £1 bn package of mechanical, electrical, heating, ventilation and air-conditioning services on the Hinkley Point C nuclear power station project
- the Hong Kong government's Drainage Services Department is planning to use it on future projects.

We suspect that this limited usage may be due to a reluctance among project sponsors to incur the inevitable time and expense that will be involved in negotiating the terms of an Alliance Contract with several different package contractors who have never been party to such an agreement before. It is perhaps difficult to justify that investment, given the availability of an alternative two-tier incentive regime based on

- an NEC4 Engineering and Construction Contract, Option C target price pain/gain share incentive within each works package contract (which would survive any termination of the project-wide incentive regime)
- an NEC4 Secondary Option Clause X12 target price gain share incentive, applied across the whole project.

Note
1. Crossrail, Thames Tideway Tunnel, Hinkley Point C and High Speed 2.

REFERENCES

ACA (Association of Consultant Architects) (2000) *The PPC Suite: Project Partnering Contracts and Alliance Forms from the ACA*. ACA, Kent, UK. https://ppc2000.co.uk/ (accessed 09/07/2024).

Cabinet Office (2014) *New Models of Construction Procurement*. Cabinet Office, London, UK. https://assets.publishing.service.gov.uk/government/uploads/system/uploads/attachment_data/file/325011/New_Models_of_Construction_Procurement_-_Introduction_to_the_Guidance_-_2_July_2014.pdf (accessed 22/05/2024).

FIDIC (Fédération Internationale des Ingénieurs-Conseils / International Federation of Consulting Engineers) (2017) *Red, Yellow and Silver Books*. FIDIC, Geneva, Switzerland. https://fidic.org/themes/new-fidic-contracts-2017-2nd-editions-red-yellow-and-silver-books (accessed 09/07/2024).

JCT (Joint Contracts Tribunal) (2016) *Contract Families*. JCT, London, UK. https://www.jctltd.co.uk/category/contract-families (accessed 090/07/2024).

Latham M (1994) *Constructing the Team*. HMSO (Her Majesty's Stationery Office), London, UK. https://constructingexcellence.org.uk/constructing-the-team-the-latham-report/ (accessed 08/07/2024).

National Audit Office (2020) *Investigation into the Rescue of Carillion's PFI Hospital Contracts*. National Audit Office, London, UK. https://www.nao.org.uk/wp-content/uploads/2020/01/Investigation-into-the-rescue-of-Carillions-PFI-hospital-contracts.pdf (accessed 22/05/2024).

NEC (New Engineering Contract) (2017) *NEC Contracts*. NEC, London, UK. https://www.neccontract.com/ (accessed 09/07/2024).

emerald PUBLISHING

ice Publishing

Michael Smith, Matthew Finn and Jon Broome
ISBN 978-1-83549-897-2
https://doi.org/10.1108/978-1-83549-894-120241002
Emerald Publishing Limited: All rights reserved

Chapter 2

Early contractor involvement models: their pros and cons and when to use them

Dr Jon Broome

Abstract

This chapter describes

- the Preferred Contractor approach (which can be done under other forms including Joint Contract Tribunal (JCT, 2016), as well as New Engineering Contract (NEC, 2017) Option A and gives historical context) including comment on the JCT preconstruction services agreement (PCSA) and some comparisons, where appropriate, to NEC Early Contractor Involvement (ECI)
- ECI with stage two as a target cost contract and its variants
  - o single Engineering and Construction Contract (ECC) Option C contract with X22
  - o Professional Service Contract (PSC) followed by ECC contract
  - o Professional Service Short Contract (PSSC) followed by ECC Option C contract.
- ECI with a Project Budget (as per Option E)
- Optimum Contractor Involvement (Option C, but tendered on an outline design which, at the end of stage one, the Contractor adopts. Adjustments made to tendered prices in stage one)
- Project Alliances, mentioning both the NEC4 Alliancing Contract and PPC 2000
- A hybrid approach: Connect Plus's model for renewals on the M25.

2.1. Introduction

This chapter starts with an overview of the origins of what has come to be known as ECI arrangements. Why? Because understanding where something comes from gives context and understanding of where and when it is likely to work or not. Consequently, the next section is a short summary of what ECI is generically, the advantages (pros) and disadvantages (cons) of a generic ECI arrangement and when to use it. Importantly, given the fashionable nature of contractual arrangements, it also considers when not to use an ECI arrangement.

The main body of this chapter considers the six main different types of ECI arrangements as they have evolved. For each, it explains what it is, what are the arrangement's advantages and disadvantages and then when it is most suitable to be used. However, it should be recognised that none of these arrangements are fixed in stone and hence can be adjusted, evolved and, in some cases, combined to suit a project's or programme's particular circumstances.

Finally, project alliances are considered. Although there are three main different types, again each with their individual pros and cons, the common feature of all three is that they typically have an early contractor phase before the main contract is entered into for detailed design and construction.

For each ECI arrangement, I try to give a real-life example to bring it to life.

2.2. The origins of early contractor involvement

This section gives an overview of the origins of what have come to be known as ECI arrangements, so that the reader can understand the context of the arrangements.

ECI can be viewed as the coming together of two contractual arrangements.

(a) The *preferred contractor* approach as pioneered by big UK commercial property developers in the 1990s is still frequently used today. Here, the Client has a good idea of what they want an asset to do, what it would 'look' and 'feel' like in terms of quality and what it should cost. And they want contractors who, having agreed a Price, deliver what is wanted for pretty much that Price without unexpected or excessive increases due to variations and claims.

From the Contractor's perspective, the big incentive is a pipeline of repeat order profitable work: so, their main incentive is to please the Client over the long term by building to the agreed quality and a realistic Price which allows for a reasonable profit. If they have 'strong-armed' the Client into too high an initial price, an intelligent Client would know this and it would affect the likelihood of future orders. Likewise, making money through excessive claims would not go down well.

It also works the other way around: if the Client unreasonably suppresses the original Prices or fights reasonable claims for post-contract changes, Contractors will be less willing to work for them or insist on higher risk amounts within any Prices.

So how does it work? Essentially, the Client develops the building definition to what has become known as RIBA Stage 3 for spatial co-ordination, whereby the use and external look and dimensions are known (and hence planning applications can be submitted) with a good idea of the internal requirements: main structural form, internal layout and use of different areas, from which technical performance requirements in terms of power, ventilation, IT requirements and so on are established.

At this point, either a number of Contractors would be asked to give an outline price and other proposals or a single Contractor is chosen predominantly on a 'taxi rank' basis – that is, it's their turn. Together, the details would be further defined to a level that the Client was happy that it was going to get what it wanted and for the Contractor to price it up accurately enough to agree a realistic Price with minimal change. This phase is typically done at the Contractor's cost, so is ultimately only recovered if the building contract goes ahead.

Having agreed the details and contract Price, a detailed Design and Build contract is signed. Historically, as this is normally for a building, it was and still is typically let under a JCT form, whereupon it becomes a 'hands-off' arrangement in that the Contractor is, so far as practicable, left to deliver what was agreed, with minimal Client involvement.

To conclude, the key advantage of this approach is that the earlier involvement of the Contractor in defining and pricing the contract significantly de-risks the contract from both parties' perspective. That means keener costs and hence a keener contract Price. It also means significantly less change and hence claims after the contract has been entered into.

(b) The rise of *target cost* contracts. Target cost contracts are a development of cost reimbursable contracts. A pure cost reimbursable contract is where the Contractor is reimbursed the actual costs that it can demonstrate are spent on a particular contract, plus a fee which covers the Contractor's margin – that is, head-office overheads and profit. This fee is normally expressed as a percentage of the actual costs spent – so the more actual costs spent the more margin the Contractor makes – or, less commonly, a fixed fee. The former is illustrated in Figure 2.1 using NEC Engineering and Construction Contract terminology.

Target cost contracts are typically used where either quality is paramount, as there is no incentive for the Contractor to cut costs or, more commonly, on time-driven projects, where the lack of time means that is not possible to de-risk the contract to arrive at a sensible Price. Consequently, under a conventional arrangement, if a Price was agreed, there would be so much change that all effort would go into agreeing the frequent changes in Price rather than managing the risk out on site.

As alluded to before, the biggest disadvantage is the lack of incentive for the Contractor to reduce actual costs. Indeed, under a percentage fee arrangement, there is a positive incentive to spend more! There are a number of things that a Client can do to mitigate this incentive: define more precisely what 'actual costs' are; define reasons for disallowing costs, ranging from outright fraud to poor practice[1]; and, if possible, only use them in repeat order situations so that the Contractor knows that if it abuses the Client's trust, it will not get future work. Finally, it can also put in place an incentive mechanism to counteract this fundamental drawback and this is where target cost contracts come in.

Figure 2.1 Illustration of how a cost reimbursable contract works using the terminology of the NEC4 Engineering and Construction Contract, Option E: cost reimbursable contract (author's own)

Cost reimbursable contract
with fee percentage

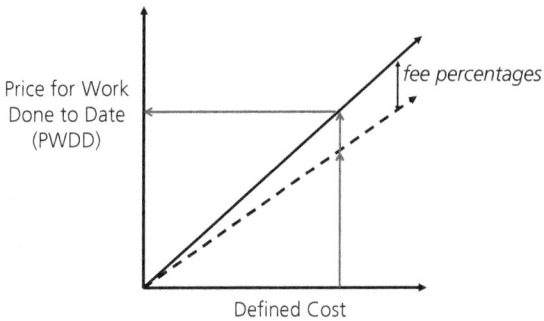

Contractually, in terms of the words in a contract, the key difference between a cost reimbursable contract and a target cost contract is that the latter has a contractually meaningful target or Prices whereby any over-run in actual costs plus fee compared with this target is shared in pre-agreed proportions and, likewise, any savings are shared in pre-agreed proportions. This is often referred to as the pain/gain share mechanism. This is illustrated in Figure 2.2 for a straight 50/50 pain/gain share.

It should be noted that the target or Price is adjusted as per a conventional price-based contract and this can be a source of contention.

Target cost contracts started to be used in the 1970s in less extreme circumstances than pure cost reimbursable contracts – that is, less risk meant that a target or Price could be agreed, albeit as an approximate estimate. However, the sharing of pain/gain meant that less risk was included in the target Price compared with a conventional lump sum contract.

CIRIA Report 85 (Perry *et al.*, 1982) made two generic observations about target cost contracts.

- Their open-book nature, where the onus is on the Contractor to show its costs in order to be reimbursed, led to openness and transparency in other areas, such as programming, risk allowances, forecasting and so on and, importantly, adjustment of the target Prices.
- The pain/gain share arrangement created an alignment of interests: it was in neither party's interests to have 'pain' and both wanted 'gain', so when a problem occurred, participants tended to work together to solve it, rather than play the 'blame game' in order to transfer liability.

Figure 2.2 Figures 2.2(a) and (b) Illustration of how a target cost contract works using the terminology of the NEC4 Engineering and Construction Contract, Option C: Target Contract with Activity Schedule (author's own)

(a) Risk sharing under a
     target cost contract

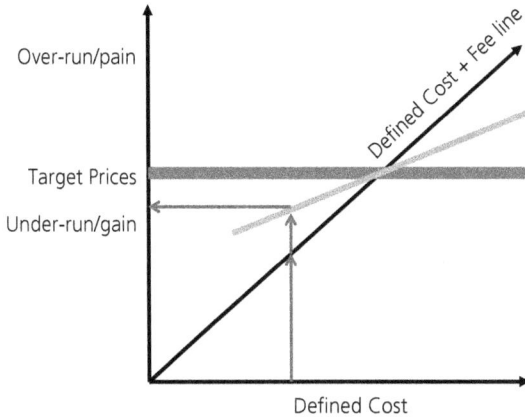

(b) Target cost contract

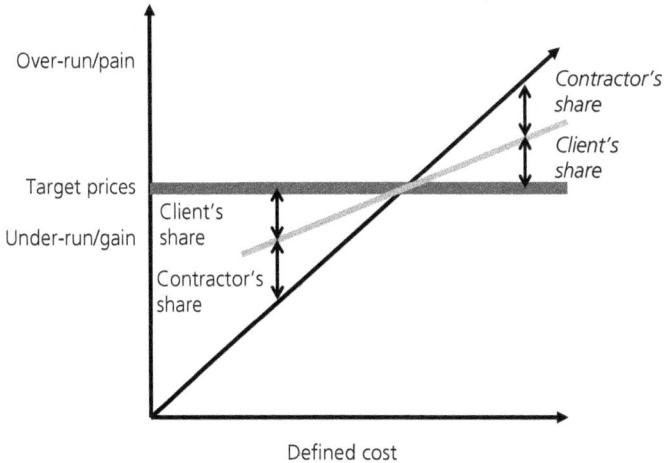

One of the main authors of the above-mentioned CIRIA report, Professor John Perry, also happened to be one of the two coinstigators of what was initially called the New Engineering Contract (now shortened to just NEC) (ICE, 1993). This had two target cost main payment options, as well as more conventional payment options (lump sum, activity schedule and bill of quantities) and a pure cost reimbursable payment option. So, for the first time, there was standard form target cost contract. As use of NEC gained in popularity throughout the 1990s (and continues to do so), the late 1990s saw the rise of partnering. And the obvious form of contract to use for partnering was NEC with a target cost option, for the reasons identified in the CIRIA report.

However, a conventionally tendered target cost contract might still mean that the target Price is too low, yet has significant unknown risk within it, so the starting point for collaboration is not where you want it to be! However, compared with a conventional lump sum or bill of quantities contract, it encourages much greater collaboration once entered into.

Bringing the features of the Preferred Contractor approach together with that of a target cost contract should, in theory, realise the advantages of both approaches: a realistic and agreed starting point in terms of scope definition and prices to enter the contract and the incentive and transparency to work together thereafter.

To my knowledge, the first big client which adopted something akin to what would now be called an ECI arrangement was the UK's Defence Estates Organisation in the late 1990s through its 'Building Down Barriers' procurement of two assets under what it called a 'prime contracting' approach. While the trials were publicised as being successful, the publicity[2] was remarkably silent on the commercial arrangement (although the NEC form was not used) and it did not seem to go anywhere afterwards – that is, the principles and practices were not expanded across the wider defence estate.

However, the ECI approach was popularised and arrived at its current name by the UK's Highways Agency (now called National Highways) which is responsible for maintaining and enhancing England's motorway and trunk road network. The Highways Agency was an early adopter of the NEC form of contract including the use of target cost options on larger contracts. The big driver for it developing an ECI approach for new roads was not actually reduced cost, but reduced time to deliver a new road, down from three parliamentary terms (i.e. up to 15 years) to within two terms (i.e. significantly less than 10 years, aiming for 8 years). The idea was that a Contractor's early involvement would ideally avoid or at least significantly shorten any planning inquiry by addressing stakeholder needs right from the outset, so that reasons for objections were either negated or significantly reduced. And in this respect, it was successful. At a high level, this approach to ECI is illustrated in Figure 2.3.

Figure 2.3 ECI as developed and initially used by the UK's Highways Agency

Overview of the ECI contracting route
e.g. original HA Early Contractor Involvement

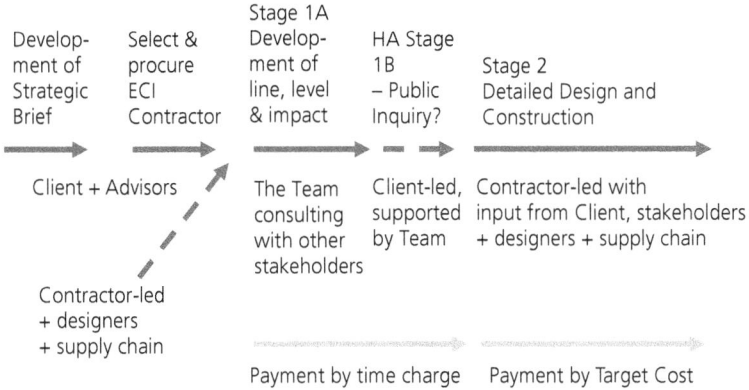

Develop-ment of Strategic Brief	Select & procure ECI Contractor	Stage 1A Develop-ment of line, level & impact	HA Stage 1B – Public Inquiry?	Stage 2 Detailed Design and Construction

Client + Advisors

Contractor-led
+ designers
+ supply chain

The Team consulting with other stakeholders

Client-led, supported by Team

Contractor-led with input from Client, stakeholders + designers + supply chain

Payment by time charge   Payment by Target Cost

2.3.   Generic ECI as we know it

Standing back from the detail of each arrangement or type (given later on in this Chapter), here are the generic characteristics of what I consider to be an ECI type arrangement.

- The Contractor is predominantly selected on quality factors as the project is not, at the point of selection, sufficiently defined for a meaningful Price to be agreed. This is not to say input costs are ignored: normally the fee percentage and professional staff rates are part of the selection criteria. For the quality part of the selection process, significant weighting is applied to the quality of the team that the Contractor is to provide and this includes its supply chain and how it will use them.

- There is a stage one in which the Contractor and its supply chain is normally paid on an open-book or time charge basis to work with the Client to develop the Scope, de-risk the project and agree a meaningful Price. The extent of Contractor involvement can vary from construction advice only, at one extreme, to leading submissions to public inquiries and developing funding submissions (as per, for example, the NHS ProCure21 and 21+ arrangements (NHS, 2024)) at the other extreme. More commonly, it involves managing the design development and doing early physical works to de-risk the construction phase (such as further site investigations based on the Contractor's specific solutions, seasonal environmental works and service diversions to take work off the critical path etc.).

▓ Following the agreement of Price and Scope, stage two is instructed for, at a minimum, construction but frequently also detailed design. Stage two is normally let under a target cost contract arrangement. However, if the Scope is well defined and there is not the need for on-going Client involvement, it could be a lump sum priced-based contract. Alternatively, the pain/gain share arrangement could be set at a higher level to include many risks which would typically be, in traditional language, a claim or variation or, in the language of the NEC contract, a compensation event. This reduces the effort in administrating change and creates further commercial alignment, but is more difficult to agree due to quantifying risk which the Contractor may only be able to influence, not control; for example, managing external stakeholders.

The generic advantages and corresponding disadvantages of a 'generic' ECI approach are listed in Table 2.1.

Table 2.1 Advantages and disadvantages of the generic ECI approach

Advantage	Disadvantage
The Contractor and its supply chain are brought in when they can add most value.	If not utilised correctly, what should be 'an investment' just becomes an additional cost.
A keener, de-risked target Price for stage two with both parties on the same page about what has to be delivered and how it will be delivered – that is, a sufficiently defined Scope with known and understood risks.	No competition to arrive at a target Price for stage two. The Client needs to be an 'intelligent client' not to be taken advantage of and ideally a repeat order one. The Contractor has more commercial leverage as deadlines approach. The negotiation may become confrontational, affecting relationships in stage two.
Less change during the contract due to the above gives both parties greater certainty.	
Easier administration of what change there is, as the Client side better understands the cost drivers due to the open-book nature of both building up the original target Price and paying on an actual cost basis.	Open-book administration, even if set up well – which it frequently isn't – is more admin heavy than a traditional lump sum priced contract.
Greater collaboration and 'value-adding' professional time to overcome issues, reduce risks and exploit opportunities that are priced within the target.	To collaborate, professionals with the right attitudes and skills need to present for both parties. Sourcing these people can be problematic. There is a danger of increased 'man-marking' as people revert to type as opposed to value-adding collaboration OR collaboration means decisions cannot be made without consensus. Both can mean bureaucracy which the Client ultimately pays for.

What struck me most when putting this table together is that most of the disadvantages stem from poor leadership/management or not having capability. This lack of capability can spring from both parties, so what does that imply?

- If you are an inexperienced Client without these capabilities, you should seriously question whether an ECI approach is right for you, even if everything about the project, taken in isolation, screams 'ECI' (see below). At a minimum, without neglecting your leadership role, you should be prepared to pay for consultancy support. And very similar selection criteria should apply to the consultant as to the Contractor!
- The selection of the right Contractor and its supply chain will be crucial to realising the benefits of ECI.
- The creation of the right culture in its broadest sense – project organisation, systems, skillsets, communication of joint interests and so on as well as the right attitudes and behaviours – will also be critical to the success.

I say more on the above, albeit packaged slightly differently, in Chapters 6 and 8.

So, when should you consider using an ECI approach? My view is when

- the project has sufficient complexity, risk and monetary value that paying for additional professional time upfront will be a value-adding investment – that is, issues will be resolved, downside risks reduced or eliminated and opportunities taken advantage of
- there is sufficient time in the project timescale before stage two – when the big money is being spent – to address the issues, risks and opportunities during stage one
- despite the investment of time in stage one, issues, risks and opportunities (which both parties can contribute to) will be still be encountered in stage two to make it worthwhile using a pain/gain arrangement for stage two to stimulate collaboration. If not, you could still take an ECI approach, but with a lump sum priced contract for stage two
- the Client and the sector in which you operate have the maturity of skills, both organisationally and individually, and behaviours to work together in a 'value-adding' way as opposed to reverting to type. This might not be overtly confrontational but could just be a man-marking 'jobsworth' manner.

2.4. The six basic ECI types and a few words on frameworks

Before I start going through the different types of ECI, I should explain that these are the six basic types as I have classified them at the time of writing. Table 2.2 provides a quick summary of the different types. For all types, stage two usually, but by no means always, includes detailed design as this may either be done by the Client or has to be virtually complete for stage two to proceed.

Table 2.2 Overview of the different types of ECI

Type	Description	Comment
1	Simple consultancy contract for advice in stage one followed by separate main contract for stage two.	Advice may include reports – for example, on approximate costs of different options.
2	Consultancy contract for professional services includes design for stage one followed by separate main contract for stage two, usually a target cost contract.	Main contract is for construction and may contain detailed design. Any minor physical works in stage one require a separate minor works construction contract.
3	One contract with stage one delivered on cost reimbursable basis and stage two on target cost contract. This is 'classic' ECI.	Break clause allows for stage two to not proceed (as for remaining types).
4	One contract with stage one delivered on cost reimbursable basis and stage two on a pain/gain basis around project budget.	Agreeing an 'all-in' price which contains what are traditionally Client risks can be problematic.
5	One contract with stage two delivered on a lump sum priced basis.	Used when reduced need for collaboration in stage two.
6	Optimum contractor involvement: target Price is fixed on entering contract rather than agreed at end of stage one.	Contractor adopts all design and more risk at end of stage one.

Not only might another type emerge but, as I stated in the introductory words to this chapter, none of these arrangements are fixed in stone and hence can be adjusted, evolved and, in some cases, combined to suit a project's or programme's particular circumstances.

And with the mention of a 'programme' of works, it is also worth stating that, for clients with a programme of works of a similar type, it will almost certainly be worthwhile and normal to enter into a framework with selected consultants and contractors. So, briefly, what do I mean by a 'framework'?

A framework is usually a longer-term agreement between a Client and one or a number of Contractors or consultants to deliver individual pieces of work under individual contracts. The duration of framework agreement is normally 4 years minimum and usually longer. The high-level framework agreement documents give the rights and obligations which sit above those stated for the individual pieces of work. This almost always includes the right of the Client not to give any work to the consultant or Contractor over the framework's duration. The framework agreement also includes or references the documents which give the procedures for how a consultant or Contractor is selected for an individual contract and how the Prices and other details are agreed, whether through a mini-competition or directly awarded.

A big advantage, particularly for public sector clients, is that the use of frameworks avoids the need to start the selection process for each contract from scratch, thus saving time and expense in procurement for all parties. Further, limited competition once selected for the framework – but not the absence of competition – means there is a repeat order incentive for each Contractor or consultant to perform to the Client's objectives rather their own short-term objectives – for example, excessive claims – as poor performance reduces the likelihood of future contracts. Finally, repeat order work allows lessons learnt from one project to be applied to subsequent ones, particularly if – and my view is that this advantage is under-utilised by clients – there is continuity of work, so that one high-performing team can move onto another project without being disbanded.

How is this relevant to ECI? Because the form of ECI that you use may be affected by whether you are in a longer-term framework for a series of similar projects or a one-off large contract.

Back to the six types, for which I will start at the simplest.

2.5. Type 1: stage one delivered under simple consultancy contract for advice only; stage two delivered under separate construction contract, typically a target cost arrangement

In this arrangement, as a Client, you want advice only, whether that is at a project definition stage (including what parts of the project will be in the subsequent contract and which parts will not be), selecting the right engineering solution to support the business case (which could include approximate costings) or providing advice to individual designers on constructability prior to them doing the design. This could take the form of being paid to attend and contribute verbally to meetings or workshops on set days; being present in a design office on a set day of the week so that designers can ask questions on preferences; or preparing option reports on the pros, cons, risks and so on of different construction options and their approximate costings. The approach is illustrated diagrammatically in Figure 2.4 and, as the caption says, could apply to type 1 and type 2 arrangements.

Because the services are relatively simple in nature, a 'simple' professional services contract, such as the NEC4 Professional Service Short Contract (PSSC) or JCT Pre-Construction Services Agreement (PCSA), can be used. Below are the generic advantages and disadvantages of this type.

Advantages
- arrangement is ideal for small/low risk/low complexity bits of work and advice
- allows really early contractor involvement to shape the project – that is, at concept and feasibility stages when maximum value can be added

Figure 2.4 Diagram summarising type 1 and type 2 ECI arrangements

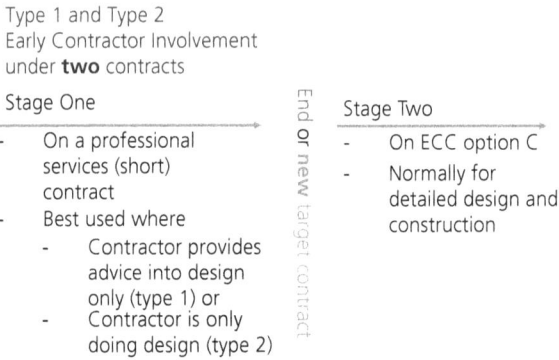

Type 1 and Type 2
Early Contractor Involvement
under **two** contracts

Stage One		Stage Two
- On a professional services (short) contract - Best used where 　- Contractor provides advice into design only (type 1) or 　- Contractor is only doing design (type 2)	End or new target contract	- On ECC option C - Normally for detailed design and construction

- allows input from a range of contractors, not just the one who will be awarded the construction contract, so encourages greater innovation early on
- particularly if a range of contractors are all given the same information and opportunities to contribute, it allows for a subsequent competition to win the main contract which could be a type 3 to 6 arrangement as well as a conventional contract
- under the NEC4 PSSC and a framework, this could be set up as a term service contract for a period of time, where each bit of consultancy is instructed as a compensation event, making administration very easy.[3]

Disadvantages

- potential duplication of ideas and work creates conflict and/or rework later on as the very early contractor involvement from one Contractor makes the project go down one technical route, but another Contractor appointed later in the process disagrees
- duplication of contracts: potentially doubles up on procurement, legal, contract admin and records. In reality, while there is some truth in this, the first contract is much simpler to administrate as a 'short' contract for advice compared with the operation of a full construction contract
- potentially precludes a single Contractor that gives the very early advice from competing for the main works contract as it has 'insider' knowledge or it locks them in
- neither the NEC4 PSSC or, to a lesser extent, the JCT PCSA are suitable for complex multidisciplinary work.

An example of type 1 ECI for advice only: Connect Plus and their COFA 3 framework

Connect Plus operate, maintain and improve the M25 London Orbital under a 30-year DBFO (design, build, finance and operate) contract on behalf of National Highways. This was awarded in May 2009. The 'operate and maintain' part of the contract, as well as the big improvements (measured in £100s of millions) in the early years of the contract, were (and continue to be) paid for on a fixed price monthly payment basis, regardless of what it costs Connect Plus or when it incurs the costs. However, since 2013, improvements up to *c.* £20 m have been funded by National Highways with the money coming through Connect Plus who, with their consultants, are responsible and accountable for their successful delivery. The previous COFA 2 (call-off framework agreement) arrangements from 2017 to 2023 contained all the contractual provisions for ECI, yet it was recognised that it was only in the later years of the 6-year framework duration that, despite having a highly collaborative culture, value from ECI started to be realised.

In its COFA 3 framework (commencing July 2023), far greater emphasis has been given in the framework documents and supporting guidance on use of ECI. As well as proposing to predominantly use type 4, with the facility to use type 3, use of type 1 ECI is available for all the tasks identified at the start of this section on type 1 ECI. In addition, type 1 will be used to work up the exact details, in terms of activities and deliverables, for stage one of a type 4 or type 3 ECI arrangement prior to entering the main NEC4 ECC. In this sense, Connect Plus is using the type 1 arrangement for 'very early' contractor involvement.[4]

How it works is that an NEC4 Professional Services Short Contract (PSSC) is entered into on an annual basis with each Contractor within the framework. The base contract will have zero work in it. However, each assignment will be instructed as a compensation event using rates for professionals which were tendered to get onto the framework. These rates are increased each year in line with construction-related inflation indices, hence the reason for an annual contract. If the assignment can be sufficiently defined, whether in hours attendance at a meeting or a specific deliverable for which a time estimate can be made, they can be priced up and paid for on a fixed cost basis once the services are completed. If not, they are paid for on a rate multiplied by actual time spent basis.

It could be argued that much or even all this service should be delivered under the framework agreement terms for 'free', rather than by way of an instructed assignment under a professional services contract. The reasons for not doing it this way are as follows.

- The extent of services of this nature is dependent on many factors, may evolve over the course of the 6-year framework and cannot be accurately forecast. For that matter, neither can the amount of physical construction. The Contractor has to recover its costs for providing the service somewhere. If not directly, it would be in its *fee percentage* for work delivered under the full NEC4 ECC and, because of the unknowns, the allowance would be an absolute guess. The COFA 3 framework documents and guidance (compared with the previous framework) are much more specific about what is to be covered by the *fee percentage* (in simple terms, limited to senior management client relationship activities) and what will be paid at tendered rates for professionals under the annual NEC4 PSSC contract (in simple terms, anything which is project-specific).
- A personal view is that Connect Plus relied too much on collaborative intent and not enough on the application of disciplined project and contract management in the previous framework. Having set out the contractual parameters and processes in more detail in the framework level documents, having to instruct and pay for specific assignments enforces a discipline on Connect Plus to define more purposely and precisely what advice it requires and any deliverables. Having done this, it puts an onus on the Contractor to do and deliver what is required.
- The Contractor is being paid to give this advice and giving good 'value-adding' advice increases the likelihood of it being allocated the main construction contract and being able to deliver it successfully. At the other extreme, if it was not being paid directly but included in its *fee percentage* for physical works, the delay between delivering the initial advice and the construction work and the possibility that these costs are not covered by the allowance in the *fee percentage* (including that the physical works contract may not actually go ahead) mean that this advice and input could be given grudgingly and hence not to the standard desired.

So, when should this arrangement be used? I would suggest when the Client
- wants (sub)contractor involvement at feasibility and concept stage in order to shape the project and main construction contract. For example, a combination of (*a*) the Client has limited understanding as it is a novel/high complexity project for the Client; (*b*) needs an idea of likely cost to see if the project is viable; (*c*) wants innovation to be 'baked in', not added in as an afterthought; (*d*) wants to ensure that the overall deal is attractive to the market
- wants specialist (sub)contractor input early on (for one or more of the above reasons) when it won't be appointing them as the main Contractor (as this does not preclude them from being a subcontractor)

▨ has an on-going relationship with the (sub)contractor: either under a framework (and it thinks it is unfair for them to do early project-specific work for no cost) or appoints them under a term arrangement using the PSSC

▨ wants to work with the chosen *Contractor* to flesh out arrangements before entering into one of the other arrangements.

Given the suggested contracts, the service required is for relatively simple tasks or just advice.

2.6. Type 2: consultancy contract for professional services including design for stage one followed by separate main contract for stage two, usually a target cost contract

In this type, the work in stage one is much meatier compared with type 1 ECI: more disciplines involved, more complex and multifaceted, with the Contractor responsible and accountable for progressing the project as opposed to merely being consulted. As well as substantial design, it may also involve stakeholder liaison, submission of planning, funding and other documents to the relevant authorities on behalf of the Client. Consequently, a more substantial consultancy contract is required which allows for this complexity. Under the NEC4 family, the full Professional Services Contract (PSC) (as opposed to the short version) would typically be used.

The big advantage of this arrangement over the following type 3 to 6 arrangements is that having two separate contracts, one for stage one and another for stage two, means there is a natural break as opposed to having to terminate or suspend the contract. This could be because

▨ based on forecast costs or other changes in external circumstances, the project is no longer viable or wanted. As any lawyer I have talked to has confirmed, it is far riskier to terminate a contract than just not sign a new one

▨ the Client no longer wishes to proceed with the same Contractor and appoints a new one. Building on the previous point, if terminating an existing contract is legally and hence commercially risky, terminating it and giving it to another contractor is even riskier!

▨ funding or other delays (such as access, planning consents etc.) may mean that stage two cannot proceed when stage one is completed, so what happens to the resource in the meantime? I heard of one project, let as a type 3, where the forecast delay was a 'best guess' but kept extending, leaving the project team in limbo with the Client paying for them. This was because the Client was unwilling to terminate as it would have to run another competition for the stage two delivery.

However, the big disadvantage compared with a type 3 arrangement is that if any physical works are required, then a consultancy-only contract is not suitable. These minor physical works could be further, more specific site investigations based on the Contractor's design solution or de-risking works, typically by taking works off the critical path and/or reducing programme interdependencies, such as

- enabling works,
- seasonal works, such as moving rare animals out of the way of the works in the off-season when they are not rearing their young,
- moving existing services out of the way before the main works begin.

In these circumstances, an additional minor works or short construction contract may need to be put in place for the Contractor to do the works. Indeed, if new requirements for works of this nature emerge at different times, multiple short or minor works contracts may need to be put in place over the duration of stage one! This can be problematic, particularly for public sector clients, if unanticipated. However, this risk can be mitigated by having management of small or specialist contractors, contracted directly with the Client under an existing framework, being part of the professional services.

Another disadvantage of the two contracts approach, particularly for one-off projects in the public sector – by which I mean those not delivered under a framework – is that the selection process would be run for two contracts which need to be linked together – that is, 'good/successful performance' on the stage one contract will lead to award of the stage two construction contract. 'Good/successful performance' (or the opposite) needs to be defined with what happens if 'bad/unsuccessful performance' occurs. This means a separate document needs to be developed which defines all this as well as what happens if minor physical works are needed.

So, when would you use a type 2 ECI arrangement?
- Compared with a type 1 arrangement, it would be when you want the Contractor to be responsible and accountable for progressing the project as opposed to just being consulted and having input.
- Compared with type 3 and other arrangements, it would be when you are less certain about the timing of the stage two main works or even whether they will actually go ahead – that is, there might be a big gap between completion of the stage one services and the stage two works. If, as a Client, I thought the same Contractor organisation would need to do limited minor physical works (of the nature discussed above) during stage one, I might recommend setting up a parallel minor works contract where physical works can be instructed as a task. However, I would be hesitant to use a type 2 arrangement if I thought any significant physical works were going to be done by the Contractor in stage one.

2.7. Type 3: 'classic' ECI: one contract with stage one on a cost reimbursable basis and stage two on a target cost basis

I say this is 'classic' ECI because it was and is the most commonly used format and, in part, this is because it is how the Addendum to the NEC3 ECC, published in 2015, and X22 of the NEC4 ECC (ECC4), published in 2017, contractually expresses it.[5] It is summarised in Figure 2.5.

Figure 2.5 'Classic' type 3 ECI under one contract

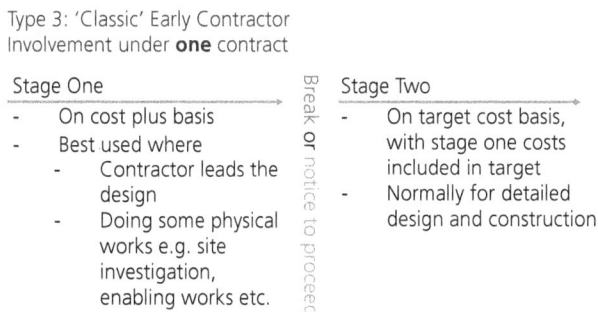

Type 3: 'Classic' Early Contractor
Involvement under **one** contract

Stage One	Break or notice to proceed	Stage Two

- On cost plus basis
- Best used where
   - Contractor leads the
     design
   - Doing some physical
     works e.g. site
     investigation,
     enabling works etc.
		- On target cost basis,
		  with stage one costs
		  included in target
		- Normally for detailed
		  design and construction

To amplify, stage one is led by the Contractor with its supply chain and, because it is typically let under a construction contract which allows for any extent of Contractor design, for example ECC3 or 4, there is no problem with the Contractor doing physical works in this stage, whether that work is further site investigation or advanced activities to de-risk the works or ensure progress of the works. However, the primary purpose of stage one is for the Contractor to lead in the development of the works so that it is sufficiently defined and de-risked for the Client to be happy with what it is to get and for a realistic target Price to be agreed for stage two.

For the works to proceed, an instruction needs to be given (under ECC by the *Project Manager* on the *Client's* behalf) whereupon the basis of payment changes from a cost reimbursable contract to a target cost contract. For stage two, the Defined Costs plus Fee (the Price for Work Done to Date (PWDD) in ECC language) from stage one are subsumed into stage two. By 'subsumed', I mean they are incorporated both into the target Prices and counted as Defined Costs already encountered.

The advantages of this approach are as follows.

- It is the industry standard arrangement, both historically and by standard conditions of contract. Consequently, as a Client, if ECI is the path you want to go down, it is the path of least resistance from the contracting community.
- Having one contract which covers early professional services, design, advanced works including further site investigations, and the main works simplifies procurement, contract administration and project management. This is a big advantage!
- The commitment of one contract by the Client 'should' provide much greater continuity of programme, key persons, Defined Costs and so on as stage one seamlessly transfers into stage two. However, notice the 'should' in the previous sentence. With the best intentions in the world, this may not happen because political, budgetary or stakeholder

issues, which are outside the control of the parties, cause delay between the completion of stage one and the notice to proceed to stage two. Having one contract may then become a disadvantage, with the Client having to pay for the Contractor's professionals to do very little or, worst case, terminating the contract and having to retender it.

▨ Related to the advantage in the previous point, the ability to do advanced works in stage one can reduce the delivery time, as well as de-risking the project by taking operations off the critical path. Consequently, for time-driven projects, this approach has significant advantages over type 2.

▨ The way the original ECC4 X22 expressed it,[6] design developed in stage two automatically becomes the Contractor's responsibility as it becomes part of the 'Scope provided by the *Contractor* for its design'. Consequently, the cost of correcting any ambiguities or inconsistencies both within it and with the Client's original Scope are shared under the pain/gain arrangement and are not compensation events leading to a change in the target Price. However, under type 2, by referencing any design developed by the Contractor in stage one as 'Scope provided by the *Contractor* for its design' from Contract Data part two, the same effect can be achieved. So arguably, this is a neutral point – neither an advantage nor disadvantage – if you know what you are doing under type 2!

The main disadvantages of this and subsequent types is that it is more likely to lead to the Contractor being commercially astute – or is that cute – at the Client's expense due to the Contractor being selected without a meaningful target Price. This can manifest itself in a number of ways.

▨ Due to the lack of competition, pricing for stage two tends to assume worst case scenarios in terms of productivity and hence duration of activities as well as allowances for risks of the sort identified in a risk register. The ability to negotiate down, or ideally genuinely mitigate through better management actions, is reduced the more the project is time-driven and as the moment of having to proceed to stage two approaches.

▨ Innovations and opportunities, including for better management of downside risk, are kept in the Contractor's back pocket, suddenly to be revealed once the notice to proceed to stage two is given and the target Price is agreed.

For both the above points, because there is less commitment to stage two under type 1 and type 2 ECI arrangements, these other arrangements encourage the Contractor to add value in stage one to curry favour and avoid any competition for stage two. In contrast, for this type 3 arrangement, the Contractor has to really over-play its hand for it not to proceed to stage two without competition.

As previously alluded to, the most common standard condition of contract which allow for this contract strategy unamended is the NEC4 ECC, main payment Option C, Target Contract with Activity Schedule and X22, ECI. However, from experience of explaining it to Clients, including their lawyers, and subsequently working with them to fine tune it to their requirements, I have the following reservations about using it unamended, although a number of these issues have subsequently been addressed in the January 2023 amendments (see the footnotes for details).

- If the *Client*, for whatever reason, decides not to proceed to stage two with the same *Contractor*, the contract does allow the *Client* to 'appoint another contractor to complete the stage two works'. However, as originally published, there were no provisions for termination of the contract expressly related to a decision not to proceed to stage two. Consequently, the general termination provisions would apply. I, and the lawyers I have worked with, much preferred an express provision to cover both this situation and if the *Client* decides not to proceed to stage two with any *Contractor*.[7]

- Nothing is said about what happens if there is a delay to stage two, in terms of the management of the *Contractor's* resources with little to do and how they are paid for. Experience suggests that they tend to drift away and be reassigned, which undermines one of the benefits of ECI. Careful thought needs to be applied to what happens here.

- The target Prices on entering the contract will be for stage one only, yet the definition of the works (as stated in the Contract Data) and Scope – that is, the technical contractual requirement – may be to deliver the full construction works, albeit expressed at a relatively high level to be developed in stage one. Does this mean the *Contractor* has an obligation to deliver the full works for the tendered stage one Prices only? Careful structuring and drafting of the Scope documents and some modifications to the words in the Contract Data are required. This becomes even more important under the type 6 ECI arrangement and I will come back to this then.[8]

- Similar logic applies to the contract Completion Date –that is, is it for the stage one works only or the whole of the completed physical works? I would say the former, in which case you would expect the contract to say a bit more than it does about how this and other dates are arrived at.

- Assuming the physical condition compensation event is not deleted, the target Prices are presumed to be based on the physical conditions which 'an experienced contractor would have judged at the Contract Date' – that is, at the start of stage one. Yet, the target Prices are set at the end of stage one when the *Contractor* might well have done more site investigations to clarify the physical conditions and based the target Prices on this risk. The date of 'judging' therefore needs to be changed to some time towards the end of stage one when the *Contractor* submits its Prices.[9]

- X22 has Pricing Information 'which specifies how the *Contractor* prepares its assessment of the Prices for stage two and is in the documents which the Contract Data states it is in' (clause X22.1(5) of ECC4). It is specified from Contract Data part two, the part the *Contractor* fills in.

  - Firstly, 'information' implies it is price or cost data filled in and used by the *Contractor* to build up the stage two Prices. And clause X22.3 (5) reinforces this impression by stating 'the total of the Prices for stage two is assessed by the *Contractor* using the Pricing Data stated in Contract Data'. However, the Contract Data part two contains other entries related to cost data. What this seems to imply is that Pricing Information is used to build up the stage two Prices, but the other cost data is used for compensation events. Why have two different sets of cost data?

○ However, reading the definition a bit more carefully, it says Pricing Information specifies 'how the *Contractor* prepares its assessment' – that is, a process, not the base data, which conflicts with how Pricing Information is used under clause X22.3 (5). As a *Client*, I would be unhappy leaving this to the *Contractor* without some direction from the *Client*. The neatest way to do this contractually would be for the *Client* to specify a high-level process in its Scope and then ask the *Contractor*, at tender and as part of the quality assessment, to develop specifics of how it would comply with the *Client's* requirements which are then incorporated into its Scope and become a contractual obligation – that is, not to have a separate Pricing Information document. And this is what I have done in the past, with all references to the Pricing Information deleted.

○ Lastly, X22.3 (1) states the *Contractor* prepares its proposals for stage two in consultation with the *Project Manager* and submits them 'in accordance with the submission procedure stated in the Scope'. Does this include Prices for stage two? If 'Yes', then we have two procedures to comply with: that stated in the Scope and that stated in the Pricing Information, with no precedence given.

I could mention some other points,[10] but standing back from the detail, the main point is that when you start to work through the details and 'what-if' scenarios, the X22 clauses as written need adding to or amending. In other words, the industry standard 'off-the-shelf' contract really needs some amending. Most of this, in my opinion, should have been done by the publishers of the contract but, regardless, the contract-specific detail still needs to be thought through and, at a minimum, specified in the Scope. It is not, therefore, an 'off-the-shelf' contract for ECI. However, perhaps with the exception of a type 1 arrangement within an ongoing framework, neither are any of the other types!

So, when should you use this arrangement?

▪ When ECI is being introduced to a new sector or country, as it is the most conventional ECI arrangement. However, in part, this presumes that the parties are already used to operating target cost contracts.

▪ Perhaps the number one requirement is that both the Client and the Contractor have the capability to constructively engage with each other to arrive at a realistic target Price for stage two. And this is not just at the tail-end of the process between the quantity surveyors: costs arise from design, construction and programming and risk allocation decisions by end-users/ultimate owners, engineers, programmers and project management people as well as commercial people. This implies both a technical capability, in terms of knowing what things should cost dependent on construction methodology, as well as a cultural aspect for all involved: traditional win–lose negotiation may result in a target Price not being agreed or, most likely, the Client, unknowingly agreeing to a high target Price. It will also affect relationships going into stage two.

Years ago, I was asked by a Contractor to evaluate some compensation events on one of the UK Highways Agency's (now National Highways) first ECI projects. With a programme of work stretching years into the future, the Highways Agency was a large repeat order client that the Contractor wanted to stay in favour with. I think those within the Highways Agency (with 20+ year memories) would admit that, at the time, they did not have the capability to evaluate the realism of the Contractor's price submission.

At the end of the contract, the Contractor had what it considered good reasons for four large compensation events. However, if it got all four, then the final Price for Work Done to Date would exceed the target Prices by more than 10%, which would make it look as if the original Prices were excessive … which the Contractor's management admitted to me was the case.

My assignment was to evaluate which two of four possible compensation events the Contractor had the best chance of getting, so that it could apply for those two only, so that the final Prices were pushed up to just below the 10%, above which the Client would start asking awkward questions!

However, unless the Client is planning to use a competition for stage two selection – which undermines the whole reason for using ECI – this criticism applies to all other types, with the exception of type 6 and, in certain circumstances, type 1.

So, when should you use this arrangement?
▪ Compared with type 1 (if not used for very early contractor involvement) and type 2, when
   o the Client is confident that stage two will proceed shortly after completion of stage one
   o the project is time-driven as construction of critical or near-critical activities can be started early in stage one.
   o Compared with type 4 to 6 arrangements, less risk is transferred to the Contractor, giving a keener monetary figure around which the pain/gain share operates. This, it could be argued, gives better value for money, especially for repeat order large clients with large projects, as they are better able to bear these risks. It also implies that agreement of the monetary figure for stage two will be easier as there is less risk to argue over in terms of its quantification and inclusion in that monetary figure. However, there are counterarguments to this, especially with type 5 arrangements, the advantages of which are explored in that section.

2.8. Type 4: stage one delivered on a cost reimbursable basis and stage two on a pain/gain basis around a project budget

The essential difference in this arrangement is that the fulcrum around which the pain/gain share operates – the contract 'Budget'[11] – is set at a higher level than the normal target Prices compared with type 3 'classic' ECI. It therefore firstly needs to include events that would normally be client risks – in NEC language, compensation events – and can also include other client costs, such as payments to statutory undertakers and landowners for use of their land while building the asset. Consequently, in stage two

- when what would otherwise be a compensation event occurs, rather than a potential fight over whether it is/is not a compensation event and, if it is, how much the target Prices should be adjusted by, the parties are now motivated to work together to resolve the matter as it is subject to pain/gain. This should result in less professionals' time spent 'administrating' the contract as there will be minimal compensation events and more time working together to 'manage in' opportunity and 'manage out' threat

- the Contractor has an interest in minimising the total Project Cost[12] to the Client, as opposed to just the construction costs for which it is traditionally responsible – for instance, on a road widening project, minimising the temporary land take alongside the road for construction purposes.

This approach can be adopted when using main payment Option E: Cost Reimbursable Contract of the NEC4 ECC, together with secondary Option X22: ECI. However, as written, it does not allow for any pain share of Project Cost above the contract Budget. The approach is illustrated in Figure 2.6.

Figure 2.6 Type 4 ECI under one contract and with gain share only for project costs below the budget

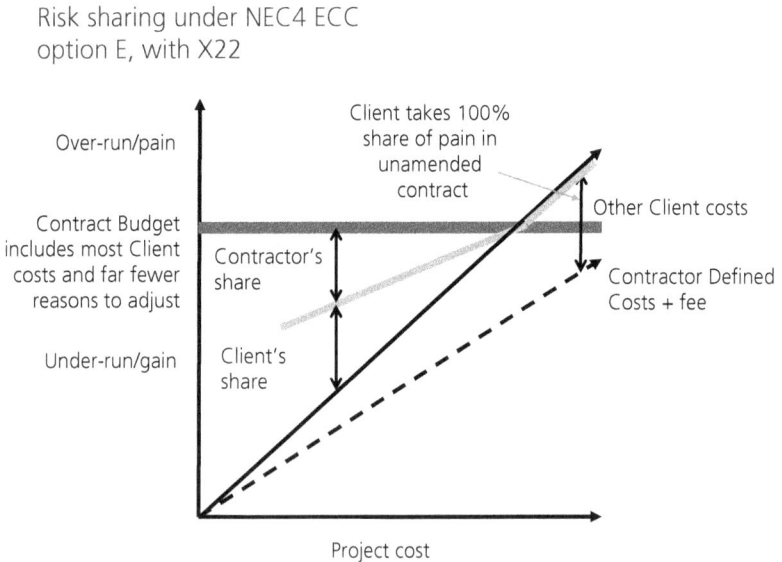

Risk sharing under NEC4 ECC
option E, with X22

Over-run/pain

Client takes 100% share of pain in unamended contract

Contract Budget includes most Client costs and far fewer reasons to adjust

Contractor's share

Other Client costs

Contractor Defined Costs + fee

Under-run/gain

Client's share

Project cost

The big initial attraction to the Client compared with other ECI types, especially if it is a large project for which they have gained external funding (whether from central Government or a financial institution), is that the funded amount can more closely match the contract Budget. However, it's when you start to think through the details that it becomes trickier!

As written, it is very attractive to Contractors as there is no pain share above the Budget, only gain share. Indeed, once the contract Budget is exceeded, as the Contractor would carry on being paid its Defined Costs (providing it can be justified and not disallowed) plus a percentage fee, its incentive is to spend more of the Client's money as it gets more fee – that is, more margin in absolute terms, including profit!

Unsurprisingly, Clients therefore want to introduce pain share if the Project Cost exceeds the contract Budget. Contractors look at this and say something like, 'so I now have a partial liability for things which are either predominantly in the control of the Client or out of either party's control' – for instance, planning approvals, statutory consents and so on. They therefore either want to
- add in risk amounts to the contract Budget, which may then exceed the funded budget. And as
  - o Contractors are generally less financially strong than Clients and
  - o Contractors look at the effect on what matters to them, which is profit, whereas Clients look at it as the effect on their overall budget for the project the amount of risk premium wanted by the Contractor seems excessive to the Client.
- have additional reasons written into the contract for adjusting the Budget (which X22 of the NEC4 ECC allows).

Both these things start to undermine the original reasons for using the type 4 arrangement. However, assuming the reasons for adjusting the Budget are limited, once it has been agreed, there is strong alignment of interests to work together to minimise the Project Cost. As with all pain/gain share arrangements, careful thought is needed as to how the pain/gain share profile works, as illustrated by the example below.

As stated for the example for type 1 ECI, Connect Plus operate, maintain and improve the M25 London Orbital under a 30-year DBFO contract on behalf of National Highways. This example is for a project jointly funded by Connect Plus and National Highways as it was part renewals and part improvement works, namely to refurbish and upgrade the two approximately 750 m-long box girder bridges, each of which carries traffic in one direction over Gade Valley in the north-west quadrant of the M25. As well as having to keep the motorway fully open, the bridge passes over the River

Gade, a main railway line and a 'B' road, both of which needed to be kept running and open. The majority of physical Works involved remedial welds, both internally and externally to the box girders. Internally meant working in confined spaces with ventilation and scaffolding access to get in. Externally involved extensive scaffolding, resulting in extensive liaison with the rail operator.

While it was known that extensive work was needed, the extent was initially unknown and involved developing innovative working methods.[13] Consequently, the early planning and physical works were awarded to Osborne Civil Engineering, now Octavius, in late 2019 under a cost reimbursable contract with the intention of converting to a target cost contract towards the end of 2020. However, the occurrence of COVID had numerous effects: procurement of goods became problematic; specialist operatives did not turn up when they should do due to the need to self-isolate; even more restrictive working practices were introduced as well as significant price fluctuation. This meant that as the end of 2020 approached – and another lockdown became imminent as COVID re-emerged – agreeing a realistic target Price became a guess and all parties recognised this. The initial reaction was to try to draft lots of additional compensation events to cover all eventualities. However, this resulted in on-going discussions over the precise wording and it was recognised that it would result in a heavy professional burden to administrate the compensation event procedure.

However, political pressures meant that a financial incentive on the Contractor was desired. Osborne (now Octavius) were also on the Connect Plus-financed renewals framework, which had a commercial relationship which included something akin to a type 4 arrangement – see the end of this chapter.

The agreed contract Budget of approximately £25.3 m had significant risk in it and allowed for a near worst-case scenario – that is, near worst-case COVID restrictions to the end of the contract. As these circumstances would be in neither party's control or influence, the Client took all the initial savings. The Contractor's share then increased to 50% within the 'realistic' range where the Contractor could be adding value and hence earning its gain share. After this, the gain share decreased because savings beyond this amount would be more due to good luck as opposed to good management. Further, the Contractor's share of savings was multiplied by a modifier which reflected other aspects of their performance.

The final Defined Cost plus Fee came in at £22.2 m, giving a saving of £3.1 m and the Contractor a gain share of £375k. This saving was largely attributed to the worst-case scenarios for COVID not occurring and good management.[14]

> The success of this arrangement led Connect Plus to use it on several additional projects at the tail-end of its COFA 2 framework for improvements. These involved a more traditional stage one which focused on consultancy services but with some investigatory and enabling works, with stage two having limited pain share for any over-run: enough to keep the Contractor motivated to minimise costs, but recognising that the Client is best able to bear the over-run. These were also a success and, as a result, it is now the favoured arrangement – as opposed to a 'classic' type 3 arrangement – for Connect Plus's COFA 3 framework for improvements.

So, when is a type 4 arrangement appropriate to be used?

- As with target cost contracts – all the previous types discussed so far – when the Client can positively influence the Contractor's Defined Costs, but also when the Contractor can positively influence the Client's Direct Costs which are included in the contract Budget. Note the word 'influence': it may well be that the Client still has the most influence of the two parties, but the Contractor has enough influence to be motivated to contribute by the gain share. The same principle can be applied for what would otherwise be compensation events: whether they occur or not is typically mainly in the influence of the Client, but the Contractor has the most influence over their impact.
- When the contract hits – or is anticipated to hit – a 'sweet spot' in terms of risk maturity for what would otherwise be compensation events and other traditionally held Client costs and risks. By this I mean that the allowance is not a complete guess. This implies that, in order to agree the risk amount, the Client has some experience of previous similar work and the Contractor has some experience of both the competence of the Client (as many compensation events arise from their actions or inactions) and the type of work. Having said this, as per the example above, the pain/gain share can be set to accommodate uncertainty.
- The above point implies some sort of on-going relationship between the parties so that they know and trust each other. This may well be within a framework contract or agreement, with a number of similar contracts being delivered over a period of time.

2.9. Type 5: one contract with stage two delivered on a lump sum priced basis

At a concept level, this works exactly as per type 3, except that stage two is paid for on a lump sum or activity schedule basis[15] – that is, the Contractor is paid what it has priced each operation or activity at, rather than a cost plus fee basis. Once the Prices have been agreed, compared with a cost plus fee pain/gain sharing arrangement, this gives both parties more certainty of what the Contractor will be paid and reduces contract administration.

However, if we take a theoretically same contract in terms of what the Contractor has to deliver and associated risk profile, it does imply that the Contractor, often the less financially strong party, will include more risk in the lump sum Prices compared with a target cost Price basis – that is, the Client won't necessarily be getting value for money for this risk transfer.

However, below are three circumstances for using a price-based, as opposed to cost plus with pain/gain share, payment mechanism for stage two.

- It undoubtedly costs more to financially administrate a cost-based contract. If the size of the contract is small, all other things being equal, the costs of setting up and administrating the contract – often referred to as open-book accounting – might be excessive compared with the overall value of the contract.
- The risks, as typically identified in the risk register that are within the Prices, as well as upside opportunities, are either
  - o relatively minor in terms of probability and impact. Consequently, the range of financial outcomes for what is included in the Prices is limited to a relatively narrow range
  - o predominantly within the control of the Contractor, with the Client-side having limited ability to contribute to their better management. Any savings that are generated are therefore principally due to the Contractor, not collaboration between the parties. Consequently, under any other ECI arrangement, any savings shared with the Client would be viewed as 'lost profit' as they are not jointly earned, and the Contractor may price accordingly – that is, have a higher target Price.

All three circumstances – or a combination of three – may outweigh the benefits of using a target cost contract (greater transparency and shared incentive to collaborate) plus the additional cost of financial administration.

But if this is the rational decision to make at a high level, it then presents some problems at the practical level of contract drafting and administration in terms of how the Contractor is paid for stage one and stage two. If it is one contract and you intend to pay the Contractor in stage one on a cost plus or time charge basis, then it means you are fundamentally changing the basis of payment midway through a contract. So, what are your options here?

- Draft substantial amendments to have one payment mechanism for stage one (cost plus fee) and another for stage two (lump sum/activity schedule). There really would have to be a good reason to do this.
- Have two separate contracts with stage one being on a Professional Services Contract. This has both the advantages and disadvantages of type 2 arrangements as, apart from the method of payment for stage two, it is the same arrangement.

■ Pay for stage one on the same basis: lump sum or activity schedule. To my knowledge, the NEC4 ECC contract is the only standard form with ECI clauses. However, the title of the secondary option is 'Option X22: Early Contractor Involvement (used only with Options C and E)'.

However, an analysis of the X22 clauses reveals only two sentences which reference payment mechanism and these are at the end of clause X22.3 (2). These clauses could be deleted and replaced with a like-for-like statement referencing Option A rather than C or E – that is, the replacement sentence would say 'The total of the Prices is in the form of revisions to the Activity Schedule'. That would be it!

However, it would mean that stage one is done on the basis of priced activity schedule. That in turn implies that the deliverables and/or activities that the Contractor is to do in stage one are more defined than under a type 3 arrangement, which is no bad thing! I have summarised this approach in the diagram in Figure 2.7.

This approach would arguably be an improvement on the Preferred Contractor approach in two respects.

■ It means that the Contractor is paid directly for the work it does in the 'Preferred' stage (equivalent to stage one) rather than having to recoup its costs under the 'Design and Build' stage (equivalent to stage two).

Figure 2.7 ECI type 5 under one contract using Option A: Priced Contract with Activity Schedule for both stage one and two

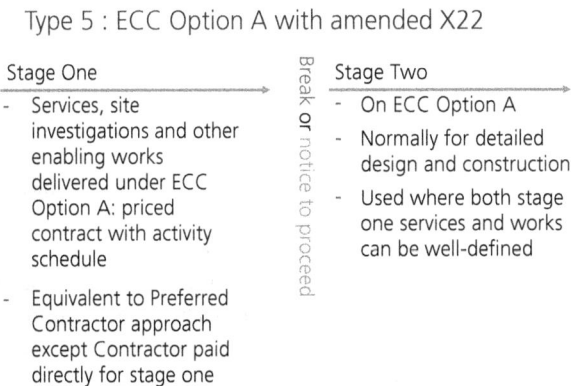

Type 5 : ECC Option A with amended X22

Stage One	Break or notice to proceed	Stage Two
- Services, site investigations and other enabling works delivered under ECC Option A: priced contract with activity schedule - Equivalent to Preferred Contractor approach except Contractor paid directly for stage one		- On ECC Option A - Normally for detailed design and construction - Used where both stage one services and works can be well-defined

▨ It forces the parties to define what the activities and deliverables from the first stage are. As I said before and re-emphasise, that is no bad thing. Indeed, it might be here that you use a combination of types: a simple consultancy contract for very early contractor involvement advice-only to define and price what is in stage one (as per the type 1 arrangement), before adopting this approach. Alternatively, you have a type 2 relationship with stage one delivered under a separate consultancy contract.

However, a drawback of repeat order relationships for works of a similar type is that, as it is not an 'open-book' arrangement, the Client does have feedback on the realism of costs, including savings made, going into subsequent contracts. Consequently, rather than the repeatable savings from previous successful contracts being incorporated into the reduced Prices of the following contracts – and therefore the Client having the benefit – the Contractor could feign no savings and keep the benefit.

I won't give any individual contract example to illustrate this approach in practice. However, two of my regular clients – one a water company and one a regional electricity company – have moved over to this relationship, and away from 'classic' type 3, in their latest frameworks using, for stage one, an NEC Professional Services Contract and, for stage two, the ECC, both Option A: Priced Contract with Activity Schedule. The reasons given were

▨ less reliance on 'collaboration' in stage two and more on defining the works properly in stage one
▨ issues around 'open-book' accounting (which I discuss more fully in Section 6.10 of Chapter 6).

Further, another Client needs the mechanical and electrical services to be renewed and upgraded in a high-profile, high-security inner city location while the building is kept operational, with the estimated value being approximately £60 m+. Following market consultation, the preference from the contracting community was very much for ECI, but with stage two being delivered as priced contract with activity schedule.

So, at the time of writing, there appears to be a definite shift towards this type.

To summarise, when would you use this approach? For me, it is when the nature of the project means that stage one of ECI can deliver sufficient certainty over scope and risk (which includes threat and opportunity) and hence the contract Prices, such that there is limited scope for the parties to have to collaborate closely to deliver it successfully in stage two. This 'limited scope' could be because of a combination of the relatively low value of the works – that is, there is just not that much risk within the Prices and what risk there is, the Contractor is the best placed to manage it, with the Client having little if any influence over it. Consequently, the best thing the Client can do is get out of the Contractor's way (which is not to say the parties don't co-operate when hopefully only minor challenges are encountered).

46

In addition, issues around open-book accounting (which I discuss in Section 6.10 in Chapter 6) seem to be a factor in increased use of type 5 ECI arrangements.

2.10. Type 6: 'optimal' ECI with tendered target price

At a high level, this works like 'classic' type 3 ECI except that Contractors tender, in competition, a target Price for stage two prior to entering into any commercial relationship. The Client therefore gets the benefit of a competitive price. So how might this work in more detail?

- A design is sufficiently developed for Contractors to tender a meaningful Price along with other quality commitments.

- The winning Contractor enters stage one on a cost plus fee basis to do further site investigation and, in part based on these findings, develop and near finalise its design. The focus is on de-risking stage two. Physical works, such as for enabling works, service diversions or seasonal environmental works, are only allowed if a strong case can be put forward for the benefits of doing them prior to stage two being instructed.

- Using NEC language, compensation events that occur in stage one adjust the tendered target Prices for stage two. By this I mean, for example, worse than expected physical and ground conditions which cause the design to be adjusted or value enhancements agreed with the Client's Project Manager, whether these enhancements are for improved functional performance or better whole life cost. In contrast, savings found in stage one, whether through value engineering the design or adopting alternative construction methodologies and sequences, do not adjust the target Prices downwards.

- On the instruction to proceed to stage two, risk transfer of the de-risked elements occurs. For example, the Contractor adopts the design in its entirety including that prepared by the Client or its consultants (which is not to say that if the Client subsequently changes this design, it isn't a compensation event) and the physical conditions compensation event is for those that could not be expected at the end of stage one, as opposed to those entering the contract – that is, the risk is transferred to the Contractor, encouraging them to de-risk the contract in stage one!

- Stage two otherwise proceeds as per a normal target cost contract.

This is illustrated in Figure 2.8.

The advantages of this approach, compared with other ECI types, are that it gives the Client

- the most price competition at tender: it therefore overcomes arguably the biggest drawback of ECI, namely lack of price competition
- greatest 'Price' certainty
    - when entering the arrangement, in that there is a contractually meaningful Price, albeit subject to adjustment by compensation events
    - arguably when going into stage two as, with the exception of type 4 with the contract Budget, there is the most risk transfer.

Type 6 : 'Optimal' ECI with tendered Price under ECC Option C with amended X22

Target prices tendered by Contractor in competition

Stage One
- Services and further site investigations delivered
- Minimal enabling works allowed
- Contractor VEs and de-risks stage two works with savings not deducted from Prices

Break or notice to proceed

Stage Two
- On ECC Option C
- Transfer of reduced risk to Contractor including responsibility for all Scope and physical conditions as known at notice to proceed

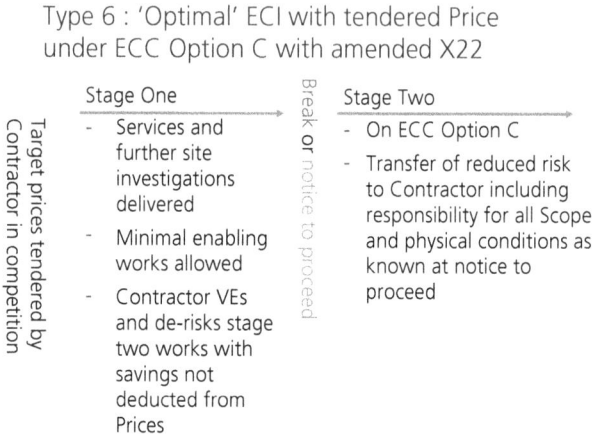

However, it does mean that, of all the different types, the design does need to be the most advanced for tenderers to price it. If too undefined, this might mean that the worse gambler is awarded the contract. If the winning Contractor then proposes significant design changes to bring the cost down to the price that it tendered or below, then that could mean significant amounts of redesign.

Another disadvantage is that it requires the most amendments to the NEC4 X22 ECI clauses and, partly because of this and its novelty, the most amount of reassurance to Contractors prior to the competition, as the example below illustrates.

The Cross Tay Link Road (CTLR) – Phase 2

The CTLR is the second phase of Perth & Kinross Council's Perth Transport Futures Project (PTFP) and is a far larger scheme than Phase 1. It will provide a new link road to the north of Perth in Scotland, UK. The PTFP consists of four phases and its main objectives are to reduce congestion and the associated pollution in Perth due to through traffic, while providing quicker journeys for those who would otherwise be going through the city. It will also enable sustainable development of the area to the north-west of the city. Each phase of the project includes measures to encourage and improve travel for non-motorised users, while Phase 4 will specifically provide measures for sustainable modes of travel within and around the city centre.

The CTLR starts just north of the city and includes a new interchange and realignment of the A9 dual carriageway. The A9 is one of, if not the, most important routes going up to the Highlands. Roundabouts are included to the west and east of the new interchange and the new link road then travels east where a new bridge takes it over the main Highland railway line and the River Tay. The bridge is a post-tensioned reinforced concrete bridge over three spans. The carriageway carries on east–west for approximately 6 km until it reaches the A94 which it joins at a roundabout junction. In-between are another two roundabouts at the A93 and Highfield. The scheme also includes a 'green bridge' and a pedestrian/cycle route over its whole length. The overall budget for the whole scheme is approximately £150.5 m, with £110.5 m funded by the Council and £40 m funded by the Scottish Government.

The value of Phase 1 was £47 m and was let as a traditional construction contract, albeit under NEC3 ECC, Option A: Priced Contract with Activity Schedule that is akin to a traditional lump sum contract. To mitigate the time impact of a procurement process, this contract was undertaken through a framework. This enabled the Council to quickly finalise the detailed design with input from the Contractor.

The optimal ECI route was adopted for the CTLR because, given the funding arrangements, the Client wanted a degree of certainty and knowledge at the start of the relationship over outturn cost. Consequently, this route was selected and developed. To go into some detail, the original Client's Scope (to use the NEC4 terminology for the technical requirements) was effectively limited to all the planning conditions, including the new road's line, level and look, with all works having to be in accordance with industry standard technical details unless a departure was approved, be that by Perth & Kinross Council for the road it would take over, Transport Scotland for the A9 or Network Rail for work in the vicinity of the railway. However, a 'non-contractual' specimen design was also supplied for the tendering Contractors which they could value engineer on to arrive at a more competitive target Price. However, whatever ideas they came up with had to comply with the Client's Scope.

Likewise, the winning Contractor could continue to value engineer this design with a requirement, at the end of a year-long stage one, for design to be near complete, otherwise the Project Manager was under no obligation to give a notice to proceed to stage two. When the notice to proceed was given, all design risk – apart from subsequent Client-instigated design changes – was within the target Prices. Further, having been given a year to develop relationships with Network Rail and Transport Scotland prior to doing any work, liaison risk was also within the Prices.

This risk transfer had to be explained and reassurance given to the contracting community. While early drafts of the contract were given to Contractors – and a lot of valuable feedback gained as a result – the Client held firm on these fundamentals and still had four out five qualifying Contractors tender, with the winning Contractor being BAM Nuttall.

At the time of writing, the contract has entered into stage two. Key points from the lessons learnt workshop for stage one included the following.

- Clearly stated requirements of what was to be delivered by when in stage two provided clarity over what was to be done, enabling good programming and delivery.
- In contrast to the original intent, bringing forwards a significant portion of the earthworks into stage one was a significant achievement which has substantially de-risked the contract. This was only possible because there was already a well-developed 'specimen' design. However, it has created some issues around subsequent design and procurement of other aspects (i.e. drainage and archaeology) which could have been anticipated.
- Having a well-developed specimen design has been very beneficial to developing third-party agreements, such as departures with Transport Scotland.
- Having the time to value engineer the design has been very beneficial, but the Contractor's processes were well-structured in this respect.

*At the time of proof reading, the contract is looking very good to be in gain and the overall project under it's budget

2.11. The three basic forms of project alliance

Firstly, what is a project alliance? The number one characteristic of a project alliance is that the incentivisation arrangement for the main players is much more aligned to the success of the overall project rather than their individual contracts. You could therefore say it is akin to a type 4 ECI arrangement, but with the big difference being that the pain/gain share applies between the Client and across the key players, rather than with just one main Contractor.

Why am I covering project alliances, albeit somewhat briefly? Because it is normal, almost universal, practice for there to be a stage one phase where

- a contractually meaningful Budget, around which the pain gain/share works, is worked up and agreed, which means …
- the technical Scope has to be worked up in sufficient, albeit high-level, detail for the Client to be confident it is going to get what it wants, and the Contractors price the Scope, including risk allowances, to give this Budget figure
- the parties get to know and trust each other including figuring out how they are going to work together to deliver the project.

In a bit more detail, in practice this means that

(a) there is a Budget set for the project which includes the Client's forecast costs as well as the key supply chain players' forecast costs
(b) this Budget includes allowances for risks and opportunities
(c) the key players, including the Client, sign up to this Budget because they can jointly influence costs, including the risks and opportunities, within this Budget
(d) costs of the key players are generally paid on an open-book or 'at-cost' basis
(e) the Budget is changed for far fewer reasons than under a conventional contract, because it includes a sufficient allowance for these risks – that is, the Budget is pretty much fixed except for significant changes in Scope and force majeure-type risks
(f) the difference between the Budget and the final costs of the key players, including the Client, is split between them in predefined portions.

Because of points (e) and (f), the parties are motivated to work together to manage in opportunity and manage out downside threat as opposed to, if point (e) did not apply, focus on pinning the blame on each other and the liability on the Client. For this arrangement to be successful, there must not only be enough risk and opportunity within the Budget (which implies that you enter a contract earlier than you would do conventionally) but that there is interdependency between the parties – that is, although different parties may be able to manage individual risks effectively, the interconnecting nature of them means it is far more efficient and effective to manage them collectively.

However, you need more than an incentive arrangement to make this work: there needs to be a degree of overarching governance, some integration of organisation and systems and some alignment of culture. I have just summarised in a sentence what is in actual fact a large commitment in order to make an alliance project successful!

There are three basic types of alliance, all with their pros and cons.

- The *traditional alliance* whereby the client signs individual contracts with each of the key supply chain partners, as illustrated in Figure 2.9 as a consultant and two contractors. These contracts are normally, but not always, cost reimbursable in nature and there is an overarching alliancing agreement, jointly signed by all key players, which ties the fortunes of the parties together for the success of the project. The contracts between the parties are black lines in Figure 2.9, while the alliancing agreement is in blue. The principal measure of success is normally outturn costs against Budget – that is, if costs are less than the Budget, the parties share in the saving and the same if there is an overrun. But there can be other measures of success, such as time to completion. Industry standard forms can be used for the individual contracts but, to my knowledge, there is no industry standard alliancing agreement which sits above the individual contracts and ties the parties' fortunes together to the overall success of the project.

Figure 2.9 Diagram illustrating scope of alliancing arrangements

Alliance contractual framework

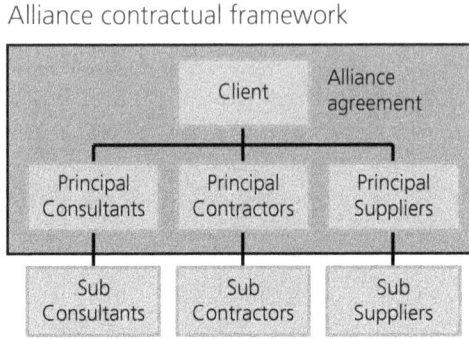

- What I call either the *simple alliance* (because it is the simplest to put in place) or an *X12 alliance* (because the NEC4 family puts it into place through use of secondary Option X12 which it calls *multiparty collaboration*). Here the *Client* signs individual contracts with each of the key supply chain partners as per a normal project. However, rather than have an overarching contract that ties the fortunes of all key parties into the success or otherwise of the overall project, X12 is added onto each individual contract. Within the X12 clauses, individual incentive payments are set against individual Key Performance Indicators. Let us say, in Figure 2.9, for every £1 that the project comes in under budget, each of the two *Contractors* get a 25% share each, while the *Consultant* gets 10%, leaving the *Client* with 40%. The reason why this is the easiest to put into place is because
  - o there is no contract between the parties to the Alliance
  - o there are no cross-liabilities – that is, there is no downside risk if another partner underperforms or mucks up in some way, so …
  - o the X12 incentive arrangements only cover upside performance.

So, what Contractor or consultant is going to refuse the opportunity to be paid more with no downside? On the other hand, Clients quite justifiably have some reservations about paying out for upside on the business case, but not getting anything back for the downside. The counterargument to this is that the *Client* does not have any more downside than if there was no X12 arrangement, while increasing the likelihood of upside. I say a lot more about the operation of X12 in my book *NEC4: A User's Guide* (Broome, 2021), but my view is that simple alliances are significantly under-used as a means of creating commercial alignment.

- The *pure alliance* or *Australian alliance*: here all the key parties sign the same contract at the same time, so they are all in it together, for better or worse. And in that sense, it is the most integrated and aligned of the three alliancing arrangements with the same arrangement and liabilities, except for quantum, applying to all its members. However, there are three main downsides to this alliancing arrangement compared with the previous two.

o   Every member has to sign the same contract at the same time, so that means all the technical and legal issues have to be agreed. This can be tricky. Indeed, I heard of one alliance contract where the contract was never actually signed by all parties because their lawyers could not agree on particular details. Yet economic necessity meant delivery started. So, what version of the 'draft' contract was the alliance actually let under?

o   A cost plus arrangement may not be feasible for some members of the alliance members. For instance, on one oil project, one of the principal suppliers was the pipeline manufacturer for the pipes which, when laid, would connect the oil field to shore. It was essential that these pipes were delivered to the timelines which suited the project – that is, if the project was going well and was 6 months in advance, as it turned out to be, they did not want to wait 6 months for the pipes to be delivered and hence the oil to start to flow. However, the vast majority of this manufacturer's costs were incurred in its factory, mixed in with all the other costs that it was incurring on other contracts – that is, it would have been near impossible to agree costs specific to the alliance contract. On both the other alliance arrangements, you can have 'fit for purpose' individual commercial and hence contractual arrangements which match the circumstances of each individual contracting party.

o   Compared with the traditional alliance and especially the simple/X12 alliance, if one partner is not performing, it can be fiendishly complicated to get them out as termination not only excludes them from being part of the alliance arrangements but also means their technical contributions are terminated as well. As a *Client*, you therefore need to find another *Contractor*, *Supplier* or *Consultant* to provide the works, goods or services, while not terminating the whole alliance!

For these reasons, as a generalisation, I prefer the traditional and simple alliance arrangements. However, as always with contracting strategy, it's horses for courses, so I would not say never use it.

In terms of UK standard forms, there is the NEC4 Alliance Contract, which includes X22: Early Alliance Involvement, and the PPC2000 contract (which stands for Project Partnering Contract with 2000 being the year it was originally published) (ACA, 2000).

The common feature of all three arrangements is that there is typically an ECI stage, hence the reason for the brief overview of each.

2.12.   Conclusions

If we include the three forms of alliancing, this chapter has outlined nine generic forms of ECI, which have the following in common.

- All have their pros and cons.
- The extent of advantages and disadvantages of each will therefore be dependent on the project and package circumstances: whether it is within a framework or not, its drivers and objectives, constraints, risk and opportunities, client and industry capabilities and so on.
- The project's and package's circumstances therefore need to be understood and defined before the 'best-fit' generic ECI type is selected.
- This can then be further refined by taking elements from other types (if appropriate and see below for an example) and lower-level decisions, as discussed in the following chapters, especially Chapter 6.

A hybrid approach by Connect Plus for renewals

As stated in previous examples, Connect Plus operate, maintain and improve the M25 London Orbital under a 30-year DBFO contract on behalf of National Highways. The example given here is for the 'Renewals' framework funded by Connect Plus as opposed to the 'Improvement' example mentioned previously, funded by National Highways.

Prior to 2017, four Contractors – two for general civils and two for pavements – each did approximately £10 m of work per annum under approximately ten separate target cost contracts each, which could range in value from a couple of hundred thousand pounds to several million. As the arrangement evolved, it trended towards assigning a package to a Contractor rather than competitive tendering as this was not found to offer value for money.

Issues with this arrangement included
- high fees as the Contractor had to recover its precontract input somewhere yet, as it was not being paid for it, was reluctant to give it freely
- difficulties in assigning costs to individual contract packages, both for Defined Costs actually incurred and in the case of compensation events
- general non-value-adding arguments over compensation events and their quantum
- a lack of incentive to collaborate across the Contractors, which is not to say they wouldn't co-operate: just that there was no commercial incentive to drive it. And given the interdependencies, it was thought that there was tremendous potential here
- missing out on savings which could be generated through continuity of work – that is, seeing the ten packages as a programme of works, rather than individual contracts.

As a result, the following arrangement was developed.
- Contractors signed an annual contract in which all ten packages were ultimately included. The pain/gain share for the individual Contractor revolves

---

around the outcomes for all packages rather than each individually. This eliminated arguments over what package a particular Defined Cost was incurred on (although data is collected for feedback purposes on what an intervention costs for future pricing and asset management purposes). It should be noted that initially only a couple of packages were mature enough to be included at the start of the year, but as the framework has progressed, more have progressively been included.

- The prices agreed are more akin to a Budget figure as per ECI type 4 arrangement – that is, they included most compensation events and some Client costs – for example, Client-procured materials. As expected, this led to some fairly robust discussions over the level of risk to be included in the Budget. However, it has meant that subsequent discussions over the quantum of compensation events have been virtually eliminated and, as data has been collected, discussions over the level of risk to be included in the Budget has become more objective as opposed to subjective. It has also demonstrably led to better management of risk by the Contractors, the savings of which are shared during the year and become the benchmark for next year's risk amounts.

- Each Contractor's direct management costs are included in the annual Budget. As packages are allocated in advance, this allows ECI for packages which will be included in the following year's contract.

- While each Contractor gets up to 30% of its own savings as gain share, 35% of the savings are effectively put into a joint pot. If one of the other Contractors makes a loss, then it draws money from this, but whatever remains is shared out between the Contractors in proportion to their final Budgets. If there is an overall deficit, then this is funded partly by Connect Plus and partly by the Contractors. In this sense, it is like an X12 simple alliance except that there is downside risk too. This really has driven collaboration, with the most common saving being sharing of lane possessions: for a typical intervention, 10% of the cost is in traffic management. So, a shared possession instantly saves 5% of cost per package.

So, has the arrangement been successful seeing as, at the time of writing (the end of 2023) it is over 6 years old? The answer is an unequivocal 'Yes', although perhaps not quite as much as hoped. It has certainly contained costs and the continuity of work and culture has made the M25 a good place to work when other high-profile projects, together with COVID and so on, could have drawn resources away. However, the proof of the arrangement is in the eating, with the Client extending the initial 6-year duration by another 4 years.

Notes

1. Under the NEC ECC target and cost reimbursable options, actual costs are defined as 'Defined Costs' and are defined in considerable detail including by reference to a detailed Schedule of Cost Components. 'Disallowed Costs' is also a defined term and defined in detail.

2. A review of the two trials and lessons learnt was the main thread of *The Handbook of Supply Chain Management: The Essentials* (Holti *et al.*, 2000).

3. Unfortunately, in my opinion, NEC4 dropped option G, Task Order option from the full Professional Services Contract (PSC) whereas the NEC3 version had it. Consequently, the NEC4 PSSC is the only unamended option for a Task Order approach for professional services and, even then, is still not a perfect fit without a few tweaks.

4. See the case study at the end of Chapter 8 for more details on precisely how this works.

5. It should be noted that, at the time of writing, the original NEC4 ECC had had three sets of amendments published. While none have been major, the last two improved the X22: Early Contractor Involvement clause, so make sure you are using the contract with the January 2023 amendments incorporated. Personally, I think there is still room for improvement, as I discuss later.

6. I use the past tense here because this was true for the original NEC4 X22 and first two amendments to it. However, in the January 2023 amendments, the contractual status of the Contractor's proposals for stage two, as accepted by the *Project Manager*, are left unstated. This needs amending to make clear!

7. Fortunately, the January 2023 amendments now cover this.

8. Ditto.

9. Again, the January 2023 amendments now cover this, albeit from the time of the *Project Manager's* notice to proceed to stage two, which may be as little as several weeks or quite a few months after the *Contractor* submits its stage two Price.

10. This is a general point which applies to ECC4 (and previous editions), regardless of whether ECI is being used or not. The contract, as written, only allows for 'Scope provided by the *Contractor* for its design' (ECC4) and not for 'Scope provided by the *Contractor* how it is to provide the works' (my words). Consequently, if the methodology for delivery is important to the *Client*, but has been developed by the *Contractor* (whether as part of the bidding process in a conventional contract or in stage one under ECI) – and the Prices have been built up on this basis – then the words of the contract need adjusting to ensure the *Contractor* delivers to this methodology – see my more detailed comments in Section 8.4 of Chapter 8.

11. 'Budget' is the terminology used in Option X22: Early *Contractor* Involvement of the NEC4 ECC.

12. Ditto. 'Project Cost' is effectively the *Contractor's* Defined Cost plus Fee plus the *Client's* other costs where a forecast allowance has been made and included in the Budget.

13. Other similar schemes around the country for Highways England involved shutting down one of the bridges and diverting the traffic onto the other bridge, with extensive traffic management for the two-way traffic. On the M25, Europe's busiest motorway, this simply was not an option due to the traffic delays it would create.

14. The full list of reasons supplied to me included

- the Contractor and supply chain always looking for efficiencies: small gains wherever we could
- consistency of the workforce – people felt part of something, more than just turning up to work
- developing a site where people wanted to work, do the right thing and behave in the right way; treating people with respect, partly driven by the Contractor's healthy employee programme
- weekly collaborative planning sessions: pro-actively identifying issues and clashes, planning around them, holding people to account and giving support where needed
- setting the expectations from the supply chain through the intelligent procurement, particularly around supervision/management/engagement with tier 1
- understanding risks and heading them off before they became reality
- upskilling the supply chain with reporting/contract administration/compliance/risk.

15. These two methods of payment are similar in that the Contractor develops its programme and then prices up the operations in that programme. The key difference is that

- under a lump sum approach (as per, for instance, the JCT Design and Build contract), the Contractor is paid, for each operation, on a percentage complete basis
- under an activity schedule basis (as per, for instance, the NEC ECC Option A: Priced Contract with Activity Schedule), the Contractor is only paid the amounts for completed activities. This tends to lead to operations being broken down into smaller activities to ensure payment. The sum of amounts for each activity form the Prices which are adjusted for exactly the same reasons, process and criteria as per the target Prices under ECC Option C: Target Contract with Activity Schedule.

REFERENCES

ACA (Association of Consultant Architects) (2000) *PPC2000: Project Partnering Contract.* ACA, London, UK.

Broome J (2021) *NEC4: A User's Guide.* ICE Publishing, London, UK.

Holti R, Nicolini D and Smalley M (2000) *The Handbook of Supply Chain Management: The Essentials.* CIRIA, London, UK.

ICE (Institution of Civil Engineers) (1993) *The New Engineering Contract.* ICE, London, UK.

JCT (Joint Contracts Tribunal) (2016) *Contract Families.* JCT, London, UK. https://www.jctltd.co.uk/category/contract-families (accessed 090/07/2024).

NEC (New Engineering Contract) (2017) *NEC Contracts.* NEC, London, UK. https://www.neccontract.com/ (accessed 09/07/2024).

NHS (National Health Service) (2024) ProCure21+. https://procure21plus.nhs.uk/ (accessed 16/07/2024).

Perry JG, Thompson PA and Wright M (1982) *Target and Cost Reimbursable Contracts: Part A: A Study of Their Use and Implications; Part B: Management and Financial Implications.* CIRIA, London, UK, Report 85.

emerald
PUBLISHING

ice
Publishing

Michael Smith, Matthew Finn and Jon Broome
ISBN 978-1-83549-897-2
https://doi.org/10.1108/978-1-83549-894-120241003

Chapter 3
Public procurement process

Catherine Maddox

Abstract

This chapter describes
- the most appropriate choice of procurement procedure for two-stage contracts
- considerations to take into account when determining the selection stage and award stage criteria
- whether or not the entire scope of the two-stage contract has in fact been procured compliantly
- the sufficiency of different pricing evaluation approaches when procuring two-stage contracts.

3.1.   Introduction

This chapter addresses the process for choosing a contractor in public procurement processes in the UK. It examines some of the challenges contracting authorities face when choosing evaluation criteria for the procurement of two-stage contracts.[1]

3.2.   Public procurement process

This section sets out the general framework of regulatory requirements applicable to a public procurement process for a public contract.

3.2.1   What is a public contract?

A public contract is a certain category of contracts awarded by a contracting authority within the meaning of the applicable public procurement legislation. For the purposes of this chapter, the Public Contracts Regulations 2015 (PCRs) are most relevant, although contracting authorities may also need to consider the applicability of the Utilities Contracts Regulations 2016 (UCRs) and the Concession Contracts Regulations 2016 (CCRs) in certain circumstances. In addition, it should be noted that the PCRs, UCRs and CCRs are due to be repealed and replaced by the Procurement Act 2023 towards the end of 2024.

3.2.2 What is a contracting authority?

Contracting authority is the term used to define the main type of purchasing entities which are required to comply with public procurement legislation. The definition includes central, regional and local authorities. It also covers a residual category of entities referred to as 'bodies governed by public law'. Whether or not a body is a body governed by public law and therefore a contracting authority is determined by applying specific criteria.[2] If a body has contracting authority status due to meeting those criteria, it is required to comply with the procedures set out in the PCRs when awarding contracts. Contracting authorities include the state, regional or local authorities and bodies governed by public law.

3.2.3 What are public procurement law obligations?

The PCRs[3] regulate the advertising and award of public contracts, which are

> contracts for pecuniary interest concluded in writing between one or more economic operators and one or more contracting authorities and having as their object the execution of works, the supply of products or the provision of services (PCR Regulation 2).

In addition, the CCRs regulate the award of works concession contracts and services concession contracts. Concession contracts are contracts for pecuniary interest concluded in writing by means of which one or more contracting authorities or utilities entrust either (*a*) the execution of works or (*b*) the provision and the management to one or more economic operators, the consideration for which consists either solely in the right to exploit the works or services that are the subject of the contract or in that right together with payment. Works concession contracts can take the form of a two-stage contract.

If a contracting authority is awarding public contracts, it is required
- to advertise the opportunity by publishing a contract notice by way of the Find a Tender Service
- to carry out one of the competitive award procedures defined in the applicable regulations
- to comply with a range of fundamental principles to ensure that bidders have the opportunity to participate in a fair and open competition for public contract opportunities. Those fundamental principles include equal and nondiscriminatory treatment of bidders and an obligation on the part of the contracting authority to act in a transparent and proportionate manner.

The public procurement regulations prescribe detailed rules and procedures in connection with the award of public contracts. Those rules cover, among other things, the criteria that may be used to shortlist bidders, rules on formulating the criteria used to award contracts, and steps that need to be taken to comply with the principles of transparency, equal treatment and nondiscrimination of bidders. For example, those rules are intended to prevent contracting authorities designing procurements with the intention of unduly favouring or disadvantaging certain bidders.

3.2.4   What happens if procurement law is breached?

Contracting authorities are under a duty to comply with the regulations and bidders can bring legal challenges, with significant consequences, where there is a breach of this statutory duty.

First, an automatic suspension on the contract making process applies where a challenge concerns an award decision. In this situation, the contracting authority cannot conclude a contract until the automatic suspension has been lifted. In most cases, this means either applying to the court for an order to lift the automatic suspension, which typically takes at least two months, or reaching an agreement with the claimant. However, in some cases the suspension can remain in place until the case is concluded, which can significantly impact the contracting authority's plans.

A range of remedies can also be ordered by the court. A successful challenge could, in principle, lead to the court ordering the contracting authority to rewind or even rerun its procurement exercise, resulting in an extensive delay to contract performance.

The courts have the power to order contracting authorities to pay damages, which are calculated on a loss of opportunity basis. Therefore, in relation to major public sector projects, damages liabilities can often be in the tens of millions of pounds.

Finally, in certain circumstances (e.g. where a contract has not been properly advertised), the court can declare a contract that has already been entered into ineffective. This means that all contractual obligations that have not yet been performed must not be performed. This remedy has a particularly draconian effect, especially in the context of works contracts, as the impact is that contractors are effectively required to 'down tools' immediately and stop work on the site as soon as a declaration of ineffectiveness is made.

3.3.   Reforming public procurement law

There are some significant upcoming changes to public procurement law, due to enter into force towards the end of 2024. For context, the procurement legislation currently in force in the UK is derived from various EU directives. In particular, the UK adopted various statutory instruments to transpose those EU directives into UK law. Following Brexit, those statutory instruments remained in force, subject to minor consequential amendments to reflect the impact of Brexit. However, the Government subsequently took the opportunity to overhaul the current public procurement regime as, following Brexit, it was no longer under an obligation to maintain that legislation in its current form.

In its 2020 Green Paper *Transforming Public Procurement* (Cabinet Office, 2020), the Government set out that it believed that the current regime is too restrictive and complex and outlined its aim to create a regulatory framework which simplifies procurement processes, placing value for money at the heart of procurement, and generating social value.

3.3.1   The Procurement Act 2023

At the time of writing, the Procurement Act 2023 (the Act) has now received royal assent and, while it introduces a number of important reforms to the current regime, many features of the new regime are substantially the same as the current regime.

The Act sets out a simpler legislative framework consisting of just one consolidated set of rules for the award of public contracts, utilities contracts, concession contracts and defence and security contracts by contracting authorities.

The new regime is expected to enter into force in October 2024. The existing rules will continue to apply for procurements started prior to the commencement date. The Procurement Regulations 2024 were laid in Parliament on 25 March 2024. They set out additional detail about various aspects of the new procurement regime.

In addition, at the time of writing, there is secondary legislation being prepared, including the draft Procurement Act 2023 (Miscellaneous Provisions) Regulations 2024 and the draft Procurement (Transparency) Regulations, which makes provision for a new online system for the publication of notices and documents that the Act requires buyers to publish, and the content of these. Many pieces of guidance are also due to be published and a wide range of Procurement Policy Notes are expected from the Cabinet Office to provide additional guidance on the new regime.

3.3.2   Current procurement procedure options

As explained above, the public procurement regulations prescribe detailed procedures which contracting authorities should use for the purposes of awarding public contracts.

The PCRs offer five alternative competitive award procedures: open, restricted, competitive procedure with negotiation (CPN), competitive dialogue (CD) and innovation partnership.

The open procedure and the restricted procedure are the default options for most procurements. However, these procedures are less likely to be appropriate for more complex opportunities where there is a need for the contracting authority to have more involved discussions and/or negotiations with bidders about the contract requirements and their proposed solutions. As such, the procurement regulations provide that contracting authorities may use the CPN, CD or innovation partnership procedures where, for example, their needs cannot be met without adaptation of readily available solutions, or their needs include design or innovative solutions, or where the contract cannot be awarded without prior negotiation because of the complexity of the contract.

Therefore, the grounds for using the CPN, CD or innovation partnership procedures will be satisfied for most two-stage contracts. In practice, the innovation partnership is a rarely used procedure and is not considered further. Therefore, the choice is typically between the CPN or CD.

3.3.3   CPN or CD?

In practice, the CPN and the CD are similar procedures. However, there are important distinctions, as shown in Figure 3.1.

Figure 3.1 Comparison of CPN and CD

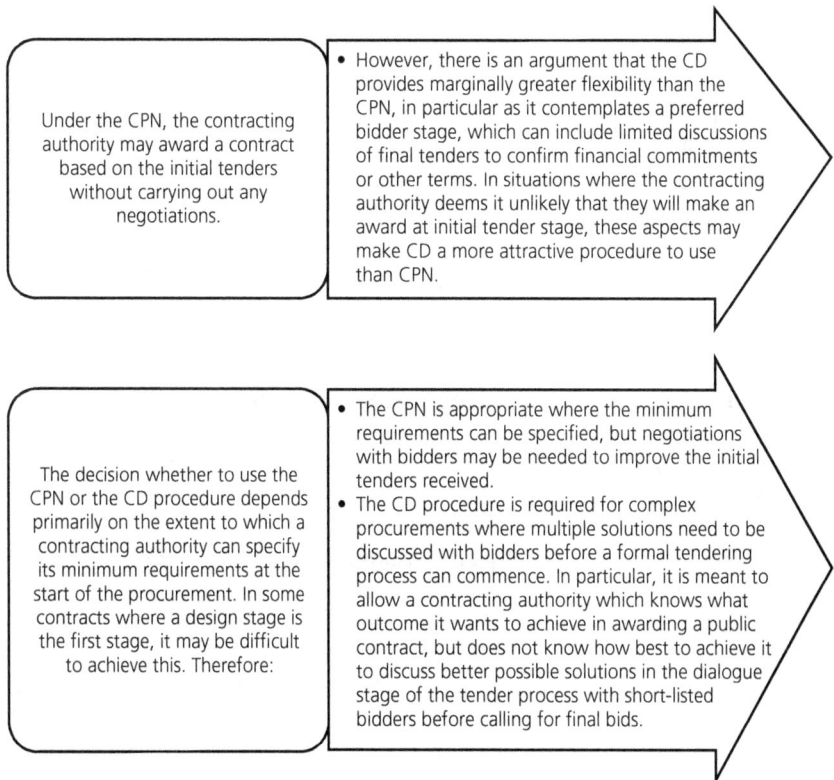

Under the CPN, the contracting authority may award a contract based on the initial tenders without carrying out any negotiations.

• However, there is an argument that the CD provides marginally greater flexibility than the CPN, in particular as it contemplates a preferred bidder stage, which can include limited discussions of final tenders to confirm financial commitments or other terms. In situations where the contracting authority deems it unlikely that they will make an award at initial tender stage, these aspects may make CD a more attractive procedure to use than CPN.

The decision whether to use the CPN or the CD procedure depends primarily on the extent to which a contracting authority can specify its minimum requirements at the start of the procurement. In some contracts where a design stage is the first stage, it may be difficult to achieve this. Therefore:

• The CPN is appropriate where the minimum requirements can be specified, but negotiations with bidders may be needed to improve the initial tenders received.
• The CD procedure is required for complex procurements where multiple solutions need to be discussed with bidders before a formal tendering process can commence. In particular, it is meant to allow a contracting authority which knows what outcome it wants to achieve in awarding a public contract, but does not know how best to achieve it to discuss better possible solutions in the dialogue stage of the tender process with short-listed bidders before calling for final bids.

3.3.4   Procurement procedure options under the Act

The Act streamlines the procedures. In particular, it contemplates only two competitive award procedures: the open procedure and the competitive flexible procedure. The competitive flexible procedure is a bespoke procedure and can incorporate various features, including successive stages, down selection and negotiation. Contracting authorities procuring two-stage contracts will therefore be using the competitive flexible procedure in most cases. In practice, contracting authorities are likely to borrow from the CPN, CD and innovation partnership when designing procurements under the competitive flexible procedure.

3.3.5   Frameworks

Although not strictly a procedure, another procurement technique that is often used is a framework. A framework agreement is an agreement between one or more contracting authorities and one or more suppliers, the purpose of which is to establish the terms governing contracts to be awarded during a given period.

It is not always appropriate to use a framework. The use of a framework will be appropriate where it is possible to establish the nature of the products to be procured and a pricing mechanism can be established. However, it is generally not appropriate to use a framework for more complex requirements, in relation to which negotiations with bidders are required prior to contract award.

However, as explained further in the case studies below, there is some evidence of contracting authorities using frameworks in the context of two-stage contract procurement as a form of resilience measure – for example, to award some of the scope of a two-stage contract to a 'reserve' bidder where the contractor exceeds the target price estimate, or to ensure that there is a degree of price competition for the stage two scope.

3.4.   How does a public procurement process work in practice?

Table 3.1 sets out the two main stages of the majority of public procurement procedures (not including the open procedure).

3.4.1   The most economically advantageous tender (MEAT)

In particular, at tender stage, the core public procurement evaluation requirement is being able to identify the MEAT, using criteria linked to the subject matter of the contract. There is also the fundamental issue of being able to identify the MEAT bid on a reliable and defensible basis – that is, choosing award criteria which are relevant to the subject matter of the contract.

The criteria used to find the MEAT almost always includes an assessment of price or cost, as set out in regulation 67 of the PCR.

Table 3.1 Selection and award stages

Stage	Description
Selection stage	A contracting authority will typically launch a public procurement process by publishing a contract notice, which advertises the opportunity. This will typically set a deadline for applicants to register interest in the opportunity and to complete a selection questionnaire (SQ) a standardised template document which must be used by all contracting authorities running above threshold public procurements. It is likely to be revised in line with the Act.
	The purpose of the SQ is to shortlist applicants to participate in the award stage.
	The SQ will include criteria and evidence requirements which are applied to identify the applicants who have the professional suitability, technical ability, relevant previous experience, capacity and financial standing to perform the contract. Typically, a mix of pass/fail and scored evaluation approaches can be used to objectively reduce the number of bidders.
Award stage	The award stage is used to assess bidders' technical and commercial solutions to deliver the contracting authority's requirements. The contracting authority is therefore required to formulate criteria so that it can assess the relative merits of bidders' solutions with a view to selecting a contractor.
	In general, award criteria should focus on proposals for performance and deliverability. Regulation 67 of the PCR sets out that contracting authorities must base the award of public contracts on the most economically advantageous tender (MEAT), identified on the basis of the price or cost, using a cost-effectiveness approach, and may include the best price/quality ratio which must be assessed on the basis of criteria linked to the subject matter of the contract. This is distinct from the approach to be taken at selection stage.
	The award stage can be iterative and may be divided into a number of phases with interim submissions and negotiations/dialogue.

(1) Contracting authorities shall base the award of public contracts on the MEAT assessed from the point of view of the contracting authority.
(2) That tender shall be identified on the basis of the price or cost, using a cost-effectiveness approach, such as life-cycle costing in accordance with regulation 68, and may include the best price–quality ratio, which shall be assessed on the basis of criteria, such as qualitative, environmental and/or social aspects, linked to the subject-matter of the public contract in question (PCR regulation 67).

Contracting authorities can and usually do also apply qualitative, environmental and social criteria, but these criteria must be proportionate and related to the subject matter of the contract.

In terms of the evaluation approach, some award criteria may be assessed on a pass/fail basis by reference to specific minimum requirements. However, there will need to be an element of scoring to identify the MEAT. Typically, the contracting authority will arrive at a price–quality ratio which it has objectively chosen in advance and modelled before finalising the exact ratio. For example, in some procurements, 60% of the overall score may be allocated to scores in response to the technical requirements and the remaining 40% may be allocated to commercial criteria (i.e. largely encompassing pricing-related criteria).

3.4.2   Award criteria in the Act

In the Act, there is a requirement for contracting authorities to award contracts on the basis of the most advantageous tender (MAT), rather than the MEAT.

The aim is to encourage broader considerations of value, including social and environmental considerations, so arguably the introduction of MAT is to encourage a slightly different point of view. However, value for money remains key and award criteria must still be linked to 'the subject matter of the contract', so that unrelated considerations are not being taken into account when evaluating the contract.

3.4.3   Award criteria when awarding two-stage contracts

The key question is how the MEAT and MAT principles can be applied in the context of two-stage contracts.

The term early contractor involvement (ECI) refers to the involvement of a contractor in the design stage of a contract – that is, prior to performance of the construction contract. Usually, when referring to two-stage tendering, the successful contractor will be awarded a preconstruction services agreement (likely to include design) (stage one) and may have an opportunity to perform a construction contract (stage two). Usually, pricing arrangements for the second stage of the contract will be determined using either a lump sum, a target cost or a guaranteed maximum price approach.

A key procurement law-related concern in relation to pricing is that typically, for a two-stage construction contract, not all elements of the design will be determined from the outset and therefore it is not practically possible to price all elements of the construction contract. This often results in a contractor being in an advantageous negotiating position in relation to agreeing the price of the second stage of the contract, as some of the pricing may be agreed bilaterally after contract award, rather than competed as part of the procurement process. Not only is this commercially undesirable, but failing to test the contractual pricing in a competitive situation can also create public procurement law risk.

A common mitigation strategy to reduce the contractor's negotiating power in this situation is to include a break clause, exercisable at the contracting authority's discretion, between the design and construction stages. This will allow the contracting authority to reprocure a different contractor to perform stage two of the contract. In theory, this may assist with introducing an element of pricing competition, if the contractor is aware that the contracting authority can walk away.

However, in practice, if the contracting authority does intend to walk away, the contracting authority will need to have built in intellectual property provisions to the contract from the outset, such as ownership over design rights, in order to make this a viable commercial approach. In addition, from a commercial perspective, there are obvious disadvantages to the contracting authority reprocuring the second stage of the contract, as the time and cost involved in reprocurement is likely to be unattractive.

As mentioned above, it is also important to consider whether or not the entire scope of the two-stage contract has in fact been procured compliantly. This will turn on a range of factors, including how the opportunity was advertised, the extent to which the evaluation truly considered the contractor's ability to deliver the entire contract scope and whether the award criteria enabled a meaningful assessment of the solutions for the entire contract scope. Government guidance on two-stage open-book processes touches on this issue (Cabinet Office, 2014).

Two-stage open-book processes involve a contracting authority inviting tender submissions based on tenderers' ability to deliver an outline brief and cost benchmark. Following the first stage competition, the appointed team works alongside the client to build up a proposal, with the construction contract being awarded at the second stage. An excerpt from the government guidance, which makes it clear that two-stage contracts can be compliantly procured in accordance with public procurement law in certain circumstances, is set out below.

(a) It is important to emphasise that inclusion of Two Stage Open Book and/or Supply Chain Collaboration in the procurement of a single project, framework, alliance or long-term contract can be compliant with EU Procurement Regulations and can be established pursuant to any EU procurement procedure (Open, Restricted, Negotiated or Competitive Dialogue).

In order to ensure that EU procurement is not an obstacle to any aspect of Two Stage Open Book or Supply Chain Collaboration, it is essential that:

(i) The Consultants and Tier 1 Contractor are appointed pursuant to an EU compliant process, either at the time of introducing Two Stage Open Book and/or Supply Chain Collaboration, or under a prior framework, alliance or long-term contract that permits Two Stage Open Book or Supply Chain Collaboration;

(ii) The criteria governing selection of the Tier 1 Contractor are appropriate to establish most economically advantageous tender and to support the Two Stage Open Book and/or Supply Chain Collaboration models; and

(iii) The Tier 1 Contractor leads the selection/review process for all Tier 2/3 Subcontractors and Suppliers in relation to their work/supply packages, and Client involvement in that selection/review process does not amount to nomination or naming of Tier 2/3 Subcontractors and Suppliers.

(b) EU Procurement and Conditional Contracts Two Stage Open Book requires the use of Conditional Contracts under which implementation of agreed preconstruction phase activities lead to Unconditional Contracts when the Client is willing to authorise commencement of construction on site. In the absence of Conditional Contracts governing Preconstruction Phase activities, a separate contract would need to be created for the Construction Phase which could break the continuity of the original Preconstruction Phase contract award and give rise to the risk of an EU procurement challenge (Using Two Stage Open Book and Supply Chain Collaboration (Cabinet Office, 2014: p.17)).

3.4.4   Finding the MEAT when procuring two-stage contracts

In many public procurements for works contracts, it is possible to scope the specification clearly and test the pricing as part of the public procurement process.

The most compliant public procurement approach for a two-stage contract is to request pricing for both stages of the contract, both the initial stage and the subsequent stage – that is, pricing for the full scope of the opportunity awarded must be competed during the procurement process. In this scenario it is possible for the contracting authority to compare bids on a straightforward basis.

However, in the case of a two-stage contract where the future specification in the second stage is not yet fully determined, there is some uncertainty as to what degree of precision in the pricing submissions is required in order to consider the contract as compliantly awarded under public procurement law.

In practice, the degree to which pricing information can be requested is dependent on the degree to which the full contract specification is capable of being understood from the outset. The difficulty with many two-stage construction contracts is that it is not possible to request meaningful pricing information for the construction stage until after the first-stage design activities have been completed. Therefore, the challenges in aligning the need for sufficient certainty as to contract scope to enable a robust pricing evaluation for procurement law purposes and the practical realities of a two-stage contract, where the design is not always known from the outset, are apparent.

If the scope and pricing of the second stage of the contract are not known at the point of bid submission, it is difficult for a contracting authority to be sure that it has identified the MEAT and adequately tested pricing during its procurement process. During the procurement process, it may be possible to easily and reliably test the total pricing for stage one, but carrying out the same exercise for stage two may present more of a challenge.

However, the procurement of two-stage contracts by contracting authorities is nonetheless relatively common and a variety of different approaches are being used to try to overcome the procurement law compliance hurdle.

3.5. General procurement law principles when evaluating pricing

Although the two-stage contracts being discussed here are construction contracts, there are no construction contract-specific public procurement rules for how the evaluation of pricing works for two-stage construction contracts. Therefore this chapter considers the issue from first principles and analogous examples. The sections below draw on industry examples of various contracts in different sectors to identify the relevant public procurement law principles that apply to evaluating pricing when publicly procuring a two-stage contract.

3.5.1 Evaluating pricing in the context of a two-stage contract

Although contracting authorities do have a wide discretion when choosing award criteria, there are some limits to this, as mentioned above. The award criteria chosen should typically include a price assessment, must be relevant to the subject matter of the contract and should be chosen in order to identify the MEAT.

Therefore, there is a question in relation to what an accurate and suitable evaluation approach may be to assess pricing proposals in the context of a two-stage contract.

3.5.2 Potential price increases

Two-stage contracts are utilised for a number of reasons, one of which is the aim of keeping costs controllable by involving the contractor early on and therefore creating an opportunity for them to assist with controlling costs.

In addition to the question about how specific and precise the pricing supplied by the contractor as part of the tender must be, there is a connected question about price increases which can arise during contract performance.

Public procurement law sets out specific rules around determining the total contract value from the commencement of the public procurement procedure (see Figure 3.2).

The contract value figure remains relevant throughout the term of the arrangement. The value figure is one of the key indicators to the market at the advertisement stage as to the scale of the opportunity. That figure is often referred to from a procurement law compliance perspective during the term of the arrangement when assessing whether modifications made to the contractual arrangements may trigger an obligation to reprocure.

Figure 3.2 Determining the total contract value

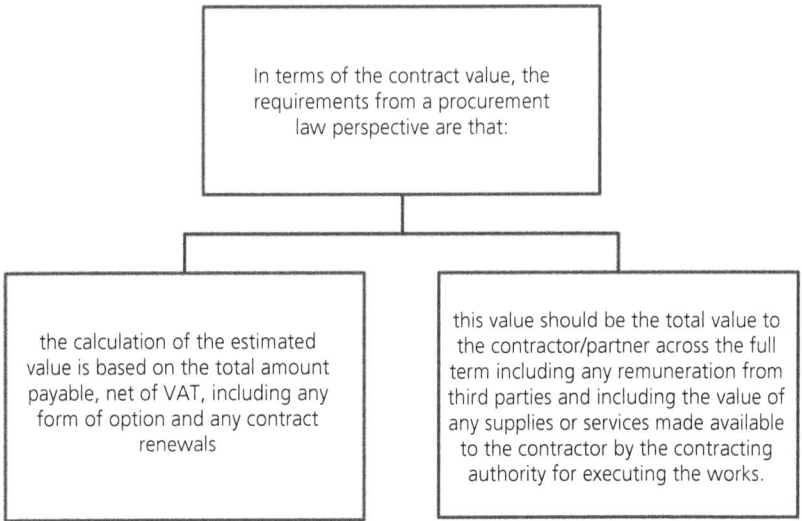

```
┌──────────────────────────────────────────┐
│  In terms of the contract value, the      │
│  requirements from a procurement          │
│  law perspective are that:                │
└──────────────────────────────────────────┘
```

| the calculation of the estimated value is based on the total amount payable, net of VAT, including any form of option and any contract renewals | this value should be the total value to the contractor/partner across the full term including any remuneration from third parties and including the value of any supplies or services made available to the contractor by the contracting authority for executing the works. |

The total contract value figure should not be exceeded during the term of the contract. The PCRs set out that the contract value may be increased without a new public procurement procedure in certain limited cases, including where the increase is less than both

- the public contract advertising threshold (at the time of writing £5 372 609 (inc. VAT) for works for central Government bodies and subcentral authorities, £139 688 (inc. VAT) for services for central Government bodies and £214 904 (inc. VAT) for services for subcentral authorities)
- 10% of the initial contract value for service and supply contracts or 15% of the initial contract value for works contracts

provided that the modification does not alter the overall nature of the contract.

Therefore, although there is some scope to exceed the total contract value, as can be seen by the requirement for any modification/increase in value to be below the advertising threshold, this scope is very minimal.

It should also be noted that there are some other grounds on which contracts can be modified, which may also in some circumstances result in an increase in contract value; however, these are not discussed in detail here (see PCR regulation 72).

With this in mind, contracting authorities need to ensure their ability to keep costs within the advertised contract value, which is likely to be a particular challenge when estimating a total contract value across a two-stage contract in advance.

One common public procurement approach is to ask contractors to price sample schemes and then apply the principles of this pricing to the pricing for the contract.

It is evidently going to be challenging, in many cases, to arrive at an accurate total cost for a two-stage contract. However, case law and case studies demonstrate that there are a range of pricing metrics which can be used, some of which are illustrated below.

3.6.   Case law example: *Henry Brothers*

The *Henry Brothers* v. *Department of Education for Northern Ireland* (2011) case is illustrative of the difficulty in selecting appropriate pricing metrics when procuring a construction contract. The court case was not in relation to a two-stage contract, but was a call off from a framework agreement. However, it is nonetheless relevant for the purposes of this analysis as it focuses on the sufficiency of the contracting authority's pricing assessment at the tender stage. The contract in question was for major school construction projects in Northern Ireland. The estimated total value of projects to be awarded under the framework was between £550 m and £650 m.

A key issue, on which the judge focused, was that the pricing evaluation was only based on a performance fee percentage structure, with no additional pricing metrics being assessed. This pricing evaluation decision was based on the assumption that costs from all bidders would be similar.

During the case, the contracting authority made it clear that it based its decision on its understanding that, in the construction market, costs did not vary significantly between contractors. Therefore, it took the view that not including an award criterion relating to cost for the construction element of the work was the best approach.

Consequently, the contracting authority established a pricing mechanism in the form of fee percentages fixed at the date of admission to the framework agreement, which would be applied later to permissible costs and thereby determine the overall cost to the contracting authority when letting contracts off the framework agreement.

The case went to trial and was then appealed, with the Court of Appeal upholding the trial judgment. The court found that the pricing mechanism in the form of fee percentages was insufficient, stating that fee percentages on their own could not determine the total price of a specific contract and that in practice, construction costs did vary, as illustrated by the excerpt from the court judgment below.

I am satisfied that the original decision to rely upon the percentage fees and bands was based upon an incorrect factual assumption sufficient to amount to a manifest error, namely, that costs would always be the same in the construction industry whether NEC option A or NEC option C was used ...

However, that is not to say, given the discretion afforded to the Department, that it is always necessary to require tenderers to carry out costing of examples or otherwise produce detailed outturn costs at the primary competition stage or that fee percentages could never be legitimately used as a pricing mechanism but in my view, as a minimum requirement, in order to comply with the Regulations and the relevant principles of community law they could only so function in conjunction with the competitive establishment of specific prices/costs at the secondary competition stage, possibly followed by a countercheck by the costs manager to safeguard against abnormally low bids etc. Otherwise the defect of using such percentages without the additional information conceded as necessary by the Department is compounded by the non-competitive establishment of specific prices/costs. The fee percentages are established in competition but the information to which they must be applied in order to function efficiently is not (*Henry Brothers* v. *Department of Education for Northern Ireland*, 2011).

The court was of the view that the contracting authority's intention to discuss rates and costs only after an individual contract had been let was insufficient to demonstrate that price had been adequately competed as part of the public procurement process. The court observed that only assessing fee percentages at final tender stage was insufficient. The court's view was that in order for the tender pricing evaluation to be sufficient, it would need to be supplemented with further rates and costs for evaluation prior to contract award.

The court found that in this context, taking that approach – that is, that rates and costs were only discussed after the contract was let – was inconsistent with the principles of transparency, equal treatment and effective competition. In addition, the court found that the assumption that costs would always be the same in the construction industry, and therefore did not need to be assessed, amounted to a manifest error.

This case illustrates the risk of challenge if a contracting authority looks to run a public procurement for major construction works but fails to provide for a sufficiently comprehensive approach for the evaluation of pricing criteria. In particular, if an insufficient range of pricing metrics are taken into account, it is difficult for the contracting authority to reach any objective determination of which tender was the most economically advantageous. Therefore, even in the context of a two-stage contract, where it can be challenging to identify suitable pricing metrics for evaluation across the full scope of the contract, an exercise to ensure that sufficient and comprehensive pricing criteria have been included in the tender documents will need to be undertaken.

3.7. Contractual options

One useful parallel to consider is what degree of pricing specificity is required when procuring a contract which includes a contractual option, as there are parallels between optional scope and the scope of the second stage of a two-stage contract.

The general public procurement law principle is to avoid further awards of work being included by way of options but not adequately tested or competed in the market. As a matter of procurement law compliance, there is a concern that contracting authorities can include option clauses in contracts as a means of circumventing their procurement law obligations. Therefore, the PCR sets out that for additional contractual performance not to be deemed as a modification which requires a new public procurement process (i.e. reprocurement), the following criteria must be met. The modifications, irrespective of their monetary value, must have been provided for in the initial procurement documents in clear, precise and unequivocal review clauses. These review clauses may include price revision clauses or options, provided that such clauses state the scope and nature of possible modifications or options as well as the conditions under which they may be used, and do not provide for modifications or options that would alter the overall nature of the contract or the framework agreement.

This arguably makes it clear that the procurement law standard for procuring a contract which includes a contractual option is largely the same standard as required for procuring the main contract. Requiring this standard avoids further awards of work being included by way of options but not adequately tested or competed in the market. This is nonetheless a fairly onerous standard.

Although selecting a suitable range of award stage pricing metrics to assess a potential contractual option can pose challenges, it is nonetheless achievable with sufficient planning at the outset of any procurement exercise. This is illustrated by the principles highlighted in the *Edenred* v. *HM Treasury* (2015) case study below, which is an example of successfully scoping in a modification clause from the outset. Contracting authorities can consider applying these principles to two-stage contracts to minimise the public procurement risk of complaints or challenges in respect of the second stage of the contract not being adequately scoped and competed during the public procurement process.

3.8. Case law example: *Edenred*

The *Edenred* v. *HM Treasury* (2015) case concerns National Savings & Investment's (NS&I) contract award to Atos to run a childcare voucher scheme. It is useful for analysing the issue of whether or not an aspect of the contract has been adequately scoped. The case covers issues such as whether the contractual option was sufficiently described in the contract notice, included in the evaluation criteria, specified precisely in the contract terms and whether the pricing for the contractual option was tested during the procurement process (see Figure 3.3).

Figure 3.3 Factors considered in the *Edenred* case

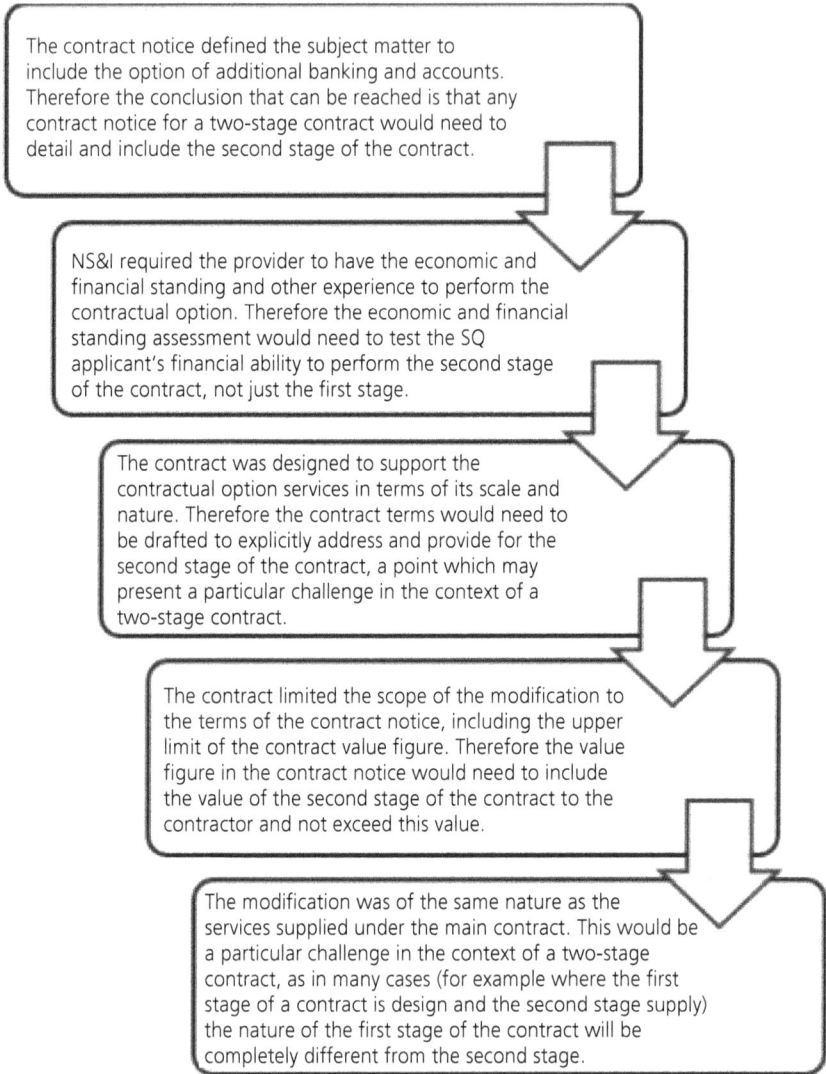

The contract notice defined the subject matter to include the option of additional banking and accounts. Therefore the conclusion that can be reached is that any contract notice for a two-stage contract would need to detail and include the second stage of the contract.

NS&I required the provider to have the economic and financial standing and other experience to perform the contractual option. Therefore the economic and financial standing assessment would need to test the SQ applicant's financial ability to perform the second stage of the contract, not just the first stage.

The contract was designed to support the contractual option services in terms of its scale and nature. Therefore the contract terms would need to be drafted to explicitly address and provide for the second stage of the contract, a point which may present a particular challenge in the context of a two-stage contract.

The contract limited the scope of the modification to the terms of the contract notice, including the upper limit of the contract value figure. Therefore the value figure in the contract notice would need to include the value of the second stage of the contract to the contractor and not exceed this value.

The modification was of the same nature as the services supplied under the main contract. This would be a particular challenge in the context of a two-stage contract, as in many cases (for example where the first stage of a contract is design and the second stage supply) the nature of the first stage of the contract will be completely different from the second stage.

In *Edenred*, the court found that the contractual option was not a modification that was substantial enough to require a new procurement process. It was therefore permitted under the procurement rules. There were a number of factors in this case that made the court conclude that the modification was not a direct award in breach of the procurement rules. Some of these factors are useful principles to apply in two-stage contracts, to maximise the likelihood that the second stage of the contract will be viewed as a compliantly competed public procurement law award from a procurement law perspective. These factors are listed above in Figure 3.3.

One of the key factors was that the contracting authority included provisions in the contract to ensure that the option did not alter the economic balance of the contract, alter the risk allocation or increase the contractor's profit margin. It is likely that in some two-stage contracts it may be possible to adopt this approach, as there could be potential to put an upper limit or cap on profit to be generated by the contractor during the second stage of the contract through a contractual mechanism.

The *Edenred* case relates to an option, rather than a two-stage contract. However, in the context of two-stage contracts, selecting a suitable range of award stage pricing metrics can be challenging; it is nonetheless achievable, as illustrated in the case study below.

3.9.   Case study example: two-stage construction contract

This case study concerns the procurement of a large-scale infrastructure project using an ECI model. An ECI model was chosen to build in design development and construction planning from early on in the process, as well as organise staff recruitment. Multiple two-stage contracts were publicly procured and the procurement was structured in four lots. The scope of the first three lots covered the main ECI contracts.

For each of the first three lots, prior to the commencement of the public procurement procedure, the specification and design was developed to the extent that the contract requirements and budget could be set and so that relevant consents could be obtained. The public procurement tender stage submissions did not require any design development, but instead focused on delivery proposals, assessing technical and commercial criteria. The commercial section of the tender submission established the stage one price and the pricing mechanism and the fees for stage two.

During stage one contract performance, the contractor undertook a range of activities, including further developing the existing design, construction planning and developing the target price. Payment in stage one was for the design and construction methodology to be developed by the contractor.

Progression to stage two was not guaranteed, but was reliant on the contractor developing and agreeing a solution with the contracting authority, agreeing a construction programme and agreeing what the target cost should be. Payment in stage two was actual costs for the construction works, plus a fee, incentivised against the target price.

In addition to the main three lots, there was also the award of a fourth lot, which was a framework of reserve contractors, a structure which was used as a resilience measure. The scope of the works to be awarded under the framework covered the scope of the works under the main contract (or the residual elements of those works) in the event of the termination or other failure of those contracts due to contractor default and other parts of the scope of the main contract in the event any element of the contract scope was omitted for any reason. Contractors appointed to the reserve framework included the successful contractors for each of the first three lots and, in the event they were different entities, the second highest scoring tenderers for each of the first three lots.

3.10.   Case study example: two-stage construction contract – analysis

Contracting authorities have a wide range of discretion as to the award criteria and metrics which they choose to use to identify the MEAT.

For example, submitting only a rate card for staff day or hourly rates for work to be done during the design stage would not reveal a sufficient amount about a bid, as not only will it not include information about the amount of time to be spent – that is, how many hours or days will be spent – but it does not address pricing for the second stage of the contract at all. Therefore, clearly further pricing metrics will need to be added. In this case, the public procurement process built in evaluation of the pricing mechanism and the fees for stage two of the contract.

Another common metric which is seen is a cap, or a bid back, on profit margin, which does add further information and can cover the second stage of the contract. However, this will need to be incorporated into the contract from the outset and again on its own will only be one element of the cost of the bid, whereas a variety of pricing metrics should ideally be evaluated.

Therefore, including a broad range of pricing assessment metrics, some of which also apply to the second stage of the contract, is advisable. The exact pricing assessment metrics chosen will be specific to the contract itself; however, it is clear from the examples that a broad range of pricing metrics, covering pricing for both stage one and stage two, should be included in the evaluation approach.

3.11.   Reserve bidder mechanisms

There is clearly precedent in the market for creating a reserve bidder mechanism or framework, as demonstrated by the case study above. Including a reserve bidder mechanism or framework is likely to add practical and commercial resilience to the project being procured.

However, depending on the exact reserve bidder mechanism used, although such a mechanism is intended to create resilience, there may be situations in which it does not operate as intended from a procurement law compliance or commercial perspective for a number of reasons.

For example, if there is an existing contractor performing one contract for one lot and then another contractor defaults and has their contract terminated and the existing contractor is requested to also perform the second contract, this may cause public procurement law difficulties. The contractor will not necessarily have had their capacity or ability to perform two contracts tested at the SQ stage. In addition, they will not have competed for and been evaluated during the tender stage of the procurement process to perform two contracts.

Therefore, in these types of scenarios, procurement law issues can arise as to whether the award of the contract has been adequately competed. With this in mind, if a reserve bidder mechanism is contemplated, the public procurement SQ and award stage evaluation criteria should be designed so that they assess the award of a contract to a reserve bidder.

Notes

1. Noting that England, Wales and Northern Ireland are due to adopt the Procurement Act 2023, although devolved Scottish authorities are excluded from its scope; therefore, Scotland will retain its current public procurement legislation, which is transposed from EU law.

2. The PCRs currently set out that bodies governed by public law have particular characteristics, which are not explored exhaustively in this chapter. In brief, they meet the following conditions. Firstly, they are established for the specific purpose of meeting needs in the general interest, not having an industrial or commercial character. Secondly, they have legal personality. Thirdly, they have any of the following characteristics: (*a*) they are financed, for the most part, by the State, regional or local authorities, or by other bodies governed by public law, (*b*) they are subject to management supervision by those authorities or bodies or (*c*) they have an administrative, managerial or supervisory board, more than half of whose members are appointed by the State, regional or local authorities, or by other bodies governed by public law. It should be noted that this definition is due to change in the Procurement Act 2023, and the concept of a public authority which is 'subject to public authority oversight' will take its place.

3. A similar definition also applies in the UCRs. The key difference is that the UCRs relate to contracts awarded by entities carrying out relevant utility activities.

REFERENCES

Cabinet Office (2014) Project Procurement and Delivery Guidance, Using Two Stage Open Book and Supply Chain Collaboration. Cabinet Office, London, UK. https://www.gov.uk/government/publications/two-stage-open-book (accessed 03/06/2024).

Cabinet Office (2020) Transforming Public Procurement. Cabinet Office, London, UK. https://assets.publishing.service.gov.uk/media/5fd77b11e90e076630958ecc/Transforming_public_procurement.pdf (accessed 20/07/2024).

Edenred v. *HM Treasury and others* (2015) UKSC 45.

Henry Brothers (Magherafelt) Ltd & Ors v. *Department of Education for Northern Ireland* (2011) NICA 59.

HMG (Her Majesty's Government) (2015) Public Contracts Regulations 2015. The Stationery Office, London, UK, Statutory Instrument 2015 No. 102. https://www.legislation.gov.uk/uksi/2015/102/contents/made (accessed 11/07/2024).

HMG (2016a) Concession Contracts Regulations 2016. The Stationery Office, London, UK, Statutory Instrument 2015 No. 273. https://www.legislation.gov.uk/uksi/2016/273/contents/made (accessed 20/07/2024).

HMG (2016b) Utilities Contracts Regulations 2016. The Stationery Office, London, UK, Statutory Instrument 2015 No. 274. https://www.legislation.gov.uk/uksi/2016/274/contents/made (accessed 20/07/2024).

HMG (His Majesty's Government) (2023) Procurement Act 2023. The Stationery Office, London, UK.

HMG (2024) Procurement Regulations 2024. The Stationery Office, London, UK, Statutory Instrument 2015 No. 692.

emerald
PUBLISHING

ice

Michael Smith, Matthew Finn and Jon Broome
ISBN 978-1-83549-897-2
https://doi.org/10.1108/978-1-83549-894-120241004

Chapter 4

Managing costs escalation on a cost reimbursable contract

Michael Smith

Abstract

This chapter
- describes both the project management tools that are available and the proactive Client engagement that is promoted within the NEC4 Engineering and Construction Contract (NEC, 2017) in order to avoid unnecessary costs escalation
- compares those provisions against other standard form contracts published by FIDIC (2017) and JCT (2016)
- describes the guaranteed maximum price, target price and incentive scheme models that have been used on some of the largest recent infrastructure projects to manage costs escalation
- explores the use of early contractor involvement and alliancing arrangements to secure best practice for the project as a whole.

4.1.　Introduction

In this chapter, we focus on the subject of cost control in the context of a cost reimbursable contract. We will look at the contractual tools that are available to the Client and its managing agent to proactively manage the risk of unnecessary costs escalation. We will also examine three different incentive schemes that have been employed recently on multi-contract project procurements in order to promote the right behaviours across the whole supply chain for the benefit of the project as a whole.

4.2.　Facilitating proactive project management

4.2.1　Project management tools

Effective costs management on a project which involves one or more cost reimbursable contracts will be largely dependent on strong and proactive management of the supply chain on behalf of the Client. So, the Client will need to give itself the necessary project management tools to discharge that critical function.

NEC4 project management provisions

The NEC4 Engineering and Construction Contract (NEC, 2017) core clauses and the Main Option C clause seek to do this by

- only allowing for the reimbursement of 'Defined Costs' and not 'Disallowed Costs', thereby encouraging the Contractor to be efficient in the management of its supply chain
- providing for a Key Dates regime, which can be used to regulate operational interface and the satisfaction of information release requirements between the various works package Contractors
- allowing the Client to prescribe specific site access arrangements within the Scope document, with which the Contractor is then obliged to comply, pursuant to core clause 25.1
- promoting programme transparency between the parties through disclosure of the Contractor's programme float and risk contingency allowances
- obliging the parties to serve early warning notices, maintain an Early Warning Register and attend early warning meetings to discuss and agree the steps necessary to mitigate the effects of the registered risks
- allowing the Project Manager to instruct a change to the Scope or a Key Date and to review and accept (or reject) the Contractor's design and programme proposals
- obliging both parties to act 'in a spirit of mutual trust and cooperation' at all times.

FIDIC project management provisions

One of the stated objectives of the second editions of the FIDIC Red, Yellow and Silver Books (FIDIC, 2017) was to enhance project management tools and mechanisms and these second editions do now embrace some, but not all, of the NEC4 concepts. The Red Book is taken here as an example.

- Clause 4.6 obliges the Contractor to accommodate works carried out by the Employer's other Contractors, public authorities and private utilities companies to the extent stated in the Specification or (as a Variation, to the extent that the instructed compliance was unforeseeable from the content of the relevant Specification provision) as otherwise instructed by the Engineer.
- A new clause 8.4 provides that both parties and the Engineer shall advise each other in advance of any known or probable future events or circumstances which may adversely affect the carrying out of any works or performance of the completed Works, increase the Contract Price or cause delay. The Engineer may then request the Contractor to submit a Variation proposal to avoid or mitigate the effect of such an event or circumstance. However, unlike NEC4, there is no express provision obliging the Engineer to take due account of any Contractor failure to so comply, in its assessment of the Contractor's entitlement in relation to that event or circumstance.

- Increased programming obligations (from 4 to 11 in total) are listed within a new clause 8.3, including disclosure of the Contractor's float and critical path.
- The Engineer is given a 21-day period to review the Contractor's initial programme submission for compliance with the contract and a 14-day period to review any revised programme (failing which the programme is deemed to comply).
- Notably, a new paragraph at the end of clause 8.5 flags up the issue of concurrent delay and allows the parties to agree the extent of the Contractor's entitlement to relief in the event of concurrent delay, by completing the Particular Provisions accordingly.

JCT project management provisions

Conversely, the JCT suite of contracts does not include a similar set of project management tools. The Construction Management Trade Contract and Construction Management Appointment, 2016 editions (JCT, 2016) are taken here as examples.

- While the Trade Contractor is obliged to comply with a programme which must evidence the critical path and be consistent with a project-wide (and Construction Manager-controlled) construction phase plan, there is no NEC4-type general obligation to revise and update the programme; only an obligation to update it to reflect extensions of time granted.
- There are no early warning or risk register maintenance provisions.
- There is no key date/activities regime and no Trade Contractor obligation to accommodate the works of others; only an entitlement to occupy so much of the site as is reasonably required.
- There is no mutual obligation on the parties to act in good faith.

4.2.2   Managing costs escalation risk on multicontract procurements

A Client about to embark on a construction management procurement should also consider whether these standard form project management tools are sufficient. In doing so, it might want to consider how package interface risk has been managed on some of the recent UK offshore wind farm procurement projects. These multi-contract, offshore procurements represent perhaps the greatest challenge in terms of managing construction interface risk. The fact that most of the recent UK offshore projects have attracted private finance is confirmation of a growing investor confidence in the ability of the leading developers in that sector to manage the significant degree of interface risk.

The key factors that these funders and investors will look at, in deciding whether they are comfortable with the project company's ability to manage interface risk, will include
- the track record of both the project sponsors and their supply chain
- clear and consistent contractual provision relating to the allocation and management of construction risk

- a detailed strategy relating to particular project delay risks (e.g. for an offshore wind farm, installation vessel availability and weather downtime strategy)
- enhanced technical adviser due diligence and modelling on the likelihood of risk occurrence
- a programme and budget which contains appropriate contingencies and which is underpinned by comprehensive insurance coverage.

The Client will also want its technical adviser to apply a top down, 'worst probable scenario' approach in its assessment of interface risk in order to determine an appropriate cost and programme contingency requirement, which is then used to inform the sizing of the base case and standby contingency funding. The technical adviser will typically request the project sponsors to provide a quantitative risk assessment in which each risk identified is quantified in terms of its probability of occurrence, its effect on cost and its effect on programme. Typically, recent projects financed in developing markets have been based on a P80 programme scenario, whereas P50 is commonly used on European projects, reflecting the more developed nature of the market.

Particularly relevant to the technical adviser's assessment will be the developer's decision to adopt parallel and/or sequential programming for the key works packages. The extent to which either approach is appropriate will largely depend on the packaging strategy that is adopted, the construction methodology and other relevant project-specific matters (e.g. weather downtime allowance and installation vessel availability, in the context of an offshore wind farm). Sequential programming is obviously likely to represent a more risk-averse approach but it may be more expensive. However, if the developer has a long-established relationship with its supply chain, which is governed by framework agreements stretching across several different projects, that should enable the developer to persuade its funders and investors that any time and cost benefits of a parallel programme are achievable.

Another key interface risk which the funder's technical adviser will wish to closely consider is the detailed design development process. The required level of design coordination will be dictated by the packaging strategy that is adopted but it will usually be administered by way of an interface matrix within each package contract which details the allocation of responsibility for such matters across the contractual nexus, an information release schedule and an accompanying 'key date' regime, pursuant to which each package Contractor is obliged to provide interface information on or before a requisite key date, failing which it will be liable to pay liquidated damages. Those liquidated damages may or may not be set at a level which keeps the developer entirely 'whole' for the consequences of the late information supply. That is something which is likely to be determined by the degree of Contractor appetite for assuming and pricing that risk within the relevant market and value for money considerations for the developer.

It is not uncommon on these projects to see the principles of supply chain collaboration and integration reflected within a master interface protocol, which might

▪ include an acknowledgement that the Contractor's rights of access to relevant parts of the site are nonexclusive in certain, specified circumstances (a concept also recognised within clause 4.6 in the FIDIC Silver Book)

▪ impose on the Contractor specific cost forecasting and programme reporting obligations

▪ oblige the Contractor to cooperate with the other package Contractors, review their programme and detailed design proposals and issue related early warning notices

▪ oblige the Contractor to populate and attend interface committee meetings in accordance with an agreed regime.

The developer may also try to secure a general right to instruct each package Contractor to accelerate its works, if that will mitigate the impact of delay on any other works package. This will inevitably lead to some debate around the associated protections (including rights to object) for the package Contractor but such problems should not be insurmountable.

4.2.3 Incentivising Contractor compliance with its project management obligations

Having given itself the contractual tools with which to manage the interface between its supply chain, the Client will want to ensure that each Contractor is incentivised to comply with its corresponding obligations. In the absence of express contractual provision setting out what the Client's remedies are in the event of noncompliance, the Client will have to place reliance on its common law right to damages for breach of contract in the relevant jurisdiction. However, it may be difficult for the Client to establish what, if any, financial loss it has incurred as a consequence of the breach and/or that such losses do not fall foul of any general limitation or exclusion of such loss that is included elsewhere in the contract. Accordingly, such a damages claim may be unlikely to act as an effective deterrent to breach of the package Contractor's project management obligations.

Accordingly, the Client may wish to negotiate the inclusion of specific remedies in the event of any Contractor failure to comply with defined project management-related obligations.

On some of the offshore wind farm projects to which we have previously referred, the following remedies have been expressly provided for.

▪ A general payment retention right until any failure to comply with the specified obligation has been remedied.

▪ The issue of management improvement notices detailing the management failure that has occurred and the improvement that is required, pending compliance with which the Client is entitled to withhold (in respect of each improvement notice served, i.e. on a cumulative basis) an amount equal to an agreed percentage from all subsequent amounts due to the Contractor.

▨ Acknowledgement by the Contractor that, if it fails to raise an early warning notice in respect of any detailed design and/or programme proposals of another package Contractor which have been provided to it for review, it will be precluded from raising a subsequent claim that the carrying out of the other Contractor's works in accordance with those proposals has caused consequent delay and/or disruption to its works. This issue is also addressed within the NEC4 Engineering and Construction Contract.

▨ Core clause 61.3 (NEC4 ECC) gives the Contractor an 8-week period in which to notify the occurrence of a compensation event, failing which (save where the event relates to the issue of an instruction or certificate by the Project Manager) the Contractor loses its right to claim in relation to that event.

▨ Pursuant to core clause 63.7, if the Contractor fails to give an early warning notice in respect of a compensation event which an experienced Contractor should have given, the Project Manager may, on notice, assess the Contractor's entitlement in respect of that compensation event as if the Contractor had given an early warning. The Contractor will therefore retain responsibility for any costs and/or delay which might have been avoided, had an early warning notice been given.

▨ Core clause 50.5 provides that, if the Contractor's programme is not included within the Contract Data, an amount equal to 25% of the Contractor's future interim payment entitlements may be retained until the programme is provided. (Notably, the Employer does not have a similar remedy in relation to the late provision of subsequent iterations of the programme.)

▨ Secondary option clause X20 (Key Performance Indicators) entitles the Employer to withhold payment of specified amounts until such time as the targets for the key performance indicators (KPI) specified in an Incentive Schedule have been met.

Related provisions within the 2017 second editions of the FIDIC suite of contracts are less comprehensive, save in relation to the Contractor's claims notification obligations. It is of particular note that, while an early warning regime has now been included, the FIDIC contracts do not replicate the NEC provision entitling the Engineer to assess the Contractor's entitlement in relation to a compensation event on the basis that an early warning notice which should have been served, was actually served.

Similar provisions are also generally absent from the JCT suite of contracts. While there is an optional KPI provision that can be utilised in the Construction Management Trade Contract and Construction Management Appointment, the Employer is mostly left only with a breach of contract claim if the Contractor fails to comply with its project management-related obligations.

4.3.   Incentive scheme models

4.3.1   Introduction

While the planning of a detailed risk management strategy, at the outset of a project, is likely to be a key element in the control of unnecessary costs escalation, the early involvement of the Client's supply chain in that planning process will be fundamental to the likelihood of its success. So, the Client will be best advised to consider how to secure the early involvement of its supply chain and, thereafter, to incentivise all members of it to work together in the best interests of the project as a whole, both during the design and planning process and during the delivery phase for the project.

We will now consider the detail of three different schemes, each involving a two-stage contract approach, based on the NEC4 ECC Option C and X22 model. Each scheme was used on a significant, recent infrastructure or power project within the UK. We have sought to highlight some of the issues that arose when the detail of those schemes was being negotiated.

4.3.2   The NEC4 Option X22 model

The NEC4 Engineering and Construction Contract provides for early contractor involvement by way of the two-stage process set out in Secondary Option X22. It enables the parties to enter into a single contract pursuant to which they develop and agree the scope of the work and the prices for it, before advancing to the construction stage.

The NEC recommend it for use with Main Option C (Target Price with Activity Schedule) or Main Option E (Cost Reimbursable Contract). When it was first published there was some suggestion that Clients wanting to engage in an ECI process when using the other main pricing options should use the NEC4 Professional Service Contract or NEC4 Professional Service Short Contract as the basis for the first stage. However, we do not see any reason why Option X22 could not be used with Main Option A (Priced Contract with Activity Schedule). That would maintain a single contract approach, thereby avoiding some of the procurement competition complications inherent in a two-contract approach. However, specialist advice should be sought in relation to the nature and extent of the amendments that would need to be made to Option X22, particularly as regards the incentive payment provisions within X22.7.

In January 2016 the NEC also published an additional Z clause for use with the NEC3 Engineering and Construction Contract, which provides for ECI. It has some small, but not insignificant, differences to the NEC4 Option X22 model. On that basis, Clients wishing to use ECI with NEC3 may wish to adopt the wording of NEC4 Option X22, making appropriate changes to align it with the NEC3 drafting style and content.

Option X22 provides for two stages, the details of which are set out by the Client in the Scope. Stage one is the preconstruction ECI phase, with development of the design and scope of the works and agreement on the pricing for the works. Stage two is the construction phase, during which any remaining detailed design is also completed.

Part two of the Contract Data specifies the Pricing Information provided by the Contractor as part of its tender and which sets out how the Contractor prepares its assessment of the Prices for stage two. If the Main Option is C (Target Contract with Activity Schedule) and the Prices for stage two are agreed at the end of stage one, the Activity Schedule is revised as appropriate and includes the Price for Work Done to Date in stage one.

At the end of stage one, the Project Manager notifies the Contractor of the Client's decision whether or not to proceed to stage two. The Client may only proceed to stage two if the Contractor has obtained the approvals and consents stated in the Scope and if the parties have agreed the total Prices for stage two, any changes to the Access Dates, the Key Dates, the Completion Date and the Budget for the Project Costs. However, the Client may decide not to proceed to stage two for any reason.

The Client's decision not to proceed to stage two is not a compensation event and it does not trigger any entitlement for the Contractor to terminate the contract. Instead, the Project Manager issues an instruction removing the stage two works from the Scope. If the Project Manager does not issue a notice to proceed to stage two because the total Prices for stage two and/or any changes to the Access Dates, the Key Dates or the Completion Date could not be agreed or because the Contractor failed to achieve the performance requirements stated in the Scope, the Client may then decide to have the stage two works performed by another Contractor.

4.3.3   The NEC4 Option X22 incentive scheme

The incentive scheme within Option X22, which is designed to drive the right behaviours during both stages one and two, is structured to operate as follows.

- A Budget is specified by the Client in part one of the Contract Data. It is the Client's estimate of the aggregate amount for the items of project cost which are included within it. If the Project Manager instructs a change to the Client's requirements stated in the Scope or if any of the other events that have been agreed by the parties (and which are specified in the Contract Data) occur, then the parties agree (or failing which, the Project Manager determines) an appropriate adjustment to the Budget.
- The Project Cost is the total paid by the Client to the Contractor and Others for the items included in the Budget.
- If, on completion of the works, the Project Cost is less than the Budget, then the Contractor is entitled to payment of the 'budget incentive', being the amount calculated by applying the percentage that is specified in the Contract Data, to the delta.

▓ An initial assessment of any budget incentive due to the Contractor is made by the Project Manager on completion of the Contractor's works and included in the next interim payment.

▓ A final assessment of any budget incentive (and any consequent balancing payment) is made when the final Project Cost has been ascertained.

This project cost-based incentive scheme can be used in addition to the Contractor's outturn cost pain/gain share incentive that is set out in the NEC4 Option C clause 54, which is examined in more detail in Section 4.3.5.

4.3.4 Points to consider when using NEC4 Option X22

The Contractor's entitlement to any budget incentive is calculated by reference to Project Cost, which is defined as the total paid by the Client to the Contractor and Others for the items included in the Budget. It is therefore important to identify, within part one of the Contract Data, what those items are – that is, the different heads of cost that are included within the Budget. By way of example, a guidance note issued by the NEC (2016) refers to the Prices, the Client's own consultancy, land purchase and other costs and any Client risk allowance. Most importantly, in order to stand the best chance of incentivising the best Contractor behaviours, the Budget should only contain items of cost which the Contractor can actually control or influence.

The user should also appreciate that the NEC scheme is based on an assumption that the Budget represents the maximum amount of money that the Client has to spend on those heads of cost that are specified within the Budget. Accordingly, pursuant to Option X22.6(1), the Budget will only be increased

▓ if the Project Manager instructs a change in any of the Client's requirements which have been included within the Scope or

▓ on the occurrence of any events that have been agreed in advance by the parties and specified in part one of the Contract Data (which might include any identified project cost risks that have not been included within the Budget, because they cannot be controlled or influenced by the Contractor, but which may influence the costs incurred by the Contractor).

The implicit reasoning is that, in these limited circumstances, the Client will have secured, or made provision for, the additional financing that is required in order to fund the increase in the Budget and can therefore preserve the budget incentive for the Contractor.

The related point to note is that Option X22 does not oblige the Client to increase the Budget to accommodate the additional cost of any compensation event that occurs during stage two (unless that compensation event is a change to the Client's requirements within the Scope or it is stated in the Contract Data as one of the events that moves the Budget). Option X22 leaves the Client free to decide whether or not to increase the Budget in such circumstances. So, the Client will perhaps need to conduct a cost–benefit analysis if a material compensation event occurs during stage two in order to inform its decision on whether or not to increase the Budget in order to preserve any potential budget incentive payment for the Contractor.

If the Client is using Option X22 together with the Main Option C pain/gain share arrangement, it should carefully consider how the two incentive schemes will work together to incentivise the Contractor to achieve savings on the Project Cost, as well as the Contractor's own construction costs. In particular

- should any Contractor entitlement to gain share on its own construction costs be payable in addition to or netted off against its entitlement to any Budget incentive payment?
- will the calibration of the two incentives be sufficient to ensure that the Contractor is incentivised to achieve a saving in Project Cost which is greater than an alternative saving in its own construction costs?

The Client's overall objective should be to ensure that the Budget incentive during stage one and the pain/gain share arrangement during stage two work together to drive the right behaviours during both stages.

The Pricing Information (or the Scope) should also set out the evaluation rules that will apply in determining any changes to the Budget. These may be the same as those which apply for the purposes of assessing compensation events, insofar as the change relates to the Contractor's construction costs, but will also need to address the impact of the event on the other elements of the Project Cost.

Case study 1

As with any two-stage procurement process, the use of Option X22 carries risk for the Employer if the process is not properly managed. The most obvious concern is that the Contractor becomes embedded in the process and, regardless of whatever the contract may say, it will be difficult for the Client to remove the Contractor from the project if the Budget, Prices and programme cannot be agreed. In other words, the consequent threat of significant programme slippage if the Contractor has to be replaced will lead to a gradual erosion of the Client's bargaining power as stage one progresses.

In this case study (which involved the use of an NEC4 Engineering and Construction Contract incorporating Main Option C and Secondary Option X22 for the procurement of a significant city centre infrastructure asset), the Client sought to manage this risk by

- setting out clear programme, work-scope and Pricing Information requirements for stage one within the Scope document included in the invitation to tender

▒ providing clearly for the Client's right to withdraw from the process without penalty at the end of stage one, to proceed with another Contractor and to use the design information provided by the Contractor (in relation to which the Client has made payment) for that purpose

▒ requiring agreement on all the Conditions of Contract that would apply in stage two as a condition precedent to the stage one appointment

▒ maintaining competitive tension within the tender process by evaluating change to the stage two Prices by reference to competitive Pricing Information submitted as part of the Contractor's tender

▒ appointing a strong design consultancy to assist the Client during the entire process with a remit to

　o investigate, analyse and develop the Client's design requirements and the Contractor's proposals

　o manage a robust value engineering process throughout

▒ incentivising the Contractor to act in the best interests of the project by entitling the Contractor to share in any saving between

　o the initial target Prices included in the Contractor's accepted tender for stages one and two

　o the aggregate of the fees paid to the Contractor for performance of the services during stage one and any revised target Prices for the stage two works which were agreed by the parties at the end of the stage one process.

Although only a percentage of any such saving was payable to the Contractor at the end of stage one, with the remainder retained by the Client until completion of the works in stage two, as security for payment of any outturn construction cost pain share for which the Contractor may then be liable.

It should be noted that the principal distinction between the incentive scheme in this case study and that which applies under Option X22 is that the Budget incentive was calculated by reference to the difference between (*a*) the Defined Cost paid to the Contractor during stage one and the revised Prices for stage two agreed by the parties at the end of stage one and (*b*) the target Prices for stages one and two included within the Contractor's accepted tender. Therefore, the pain/gain share incentive relating to the revised Prices for stage two was the only incentive scheme that applied during stage two.

Case study 2

In this case study, which involved the procurement of a major infrastructure asset within the transport sector, the Client used an NEC3 Engineering and Construction Contract, incorporating Main Option A for stage one and Main Option C for stage two, with an overarching agreement, in which it set out how the two-stage process was to operate and what the corresponding incentives for the Contractor were. Although those provisions were bespoke in nature, they were largely consistent with the provisions of NEC4 Option X22. The most noteworthy components of this contract model were as follows.

- A lump sum Price was agreed for the performance of the stage one services during a defined period. This lump sum price was £[X] less than the Contractor's tendered Price for stage one (the 'stage one Tendered Fee Excess').
- The Client's Budget for stage two could be increased at the Client's discretion but could not otherwise be changed, unless the Project Manager instructed a change to the Client's technical and/or design requirements within the Works Information or the carrying out of additional, out of scope, permanent physical works.
- The Client could terminate the contract at will at any time prior to submission of the Contractor's final proposals for stage two. Its only liability on any such termination would be to make payment on account of the stage one services performed prior to the date of termination.
- The Client was under no obligation to proceed to stage two. However, if it chose not to do so, after submission of the Contractor's final proposals for stage two, it would be liable for any breakage costs (but not loss of profit) for which the Contractor was liable under any subcontracts previously entered into with the Client's consent.
- On any termination of the Contract prior to stage two, the Contractor was obliged, at the Client's request, to novate any subcontracts to the Client.
- If a notice to proceed to stage two was issued, the Contractor was to be paid (*a*) the stage one Tendered Fee Excess and (*b*) an agreed percentage of the amount by which the aggregate of the target Prices agreed for stage two and the stage one Tendered Fee Excess was less than the Client's Budget for stage two. However, if that gain share entitlement exceeded a threshold amount, the excess amount was to be retained by the Client as security for payment of any Contractor pain share on the outturn cost of stage two.
- The Contractor's stage two pain share (payable if the Price for Work Done to Date during stage two exceeded the Client's Budget for stage two) was fixed at 100% of the excess, subject to a maximum amount equal to any Contractor stage one gain share. However, the Contractor's fee percentage entitlement was then reduced on all Defined Costs in excess of that amount.
- The invitation to tender made clear that, in submitting a tender for the contract, the tenderer would also be submitting a tender for a place on a framework agreement and that all unsuccessful tenderers would be obliged to enter into that framework agreement on contract award. That agreement sought to regulate the procedure by which a replacement Contractor would be selected to complete the contract works if the Client decided not to proceed to stage two with the original Contractor or to terminate the original Contractor's appointment at any point in time.

Case study 3

This case study involved the two-stage appointment of a process plant technology provider to develop the design of its technology solution during stage one and then manufacture, supply, install and commission the plant during stage two. The Contractor was to be paid, during both stages, on a cost reimbursable basis.

During stage one the Contractor was to develop and submits its proposals for stage two, including forecasts of the effect of those proposals on the overall project cost and programme. The Contractor's performance during stage one (i.e. at its most basic level, the delivery of acceptable proposals for stage two) was incentivised by way of the overheads and profit payment incentivisation regime referred to below and periodically assessed during a specified number of gateway reviews. These gateway reviews were designed to provide the Client with the periodic means to review and confirm whether, as its design solution developed during stage one, the Contractor was continuing to demonstrate the financial viability of the project during stage two.

Various means were employed in an endeavour to promote a collaborative working environment and incentivise Contractor performance throughout the project.

- The Client waived the right to claim liquidated damages for delay (save in relation to a key date regime in stage two which was to be used to regulate the interface between different package Contractors if they did not sign up to an alliance agreement).
- The Contractor's overheads and profit entitlements were fixed at the outset of each stage as a lump sum, calculated by applying an agreed percentage (bid by the tenderers) to the forecast cost of the relevant stage.
- The overheads entitlement fell due for payment in instalments on the achievement of specified programme milestones.
- The profit entitlement was split between payments that fell due for payment on the achievement of specified programme milestones and payments that were at risk of being moderated/reduced, depending on the level of Contractor performance ('At Risk Profit').
- If payment of At Risk Profit was reduced, the Client had the discretion to apply the whole or any part of such deduction to a new incentive scheme, to deliver project outcomes for the mutual benefit of both parties.
- If the Contractor submitted its final proposals for stage two on time and the Client decided to proceed to stage two, the Contractor became entitled to a success fee, calculated by applying a multiplier to the stage one At Risk Profit. Payment of part of that success fee was at risk of being moderated, depending on the Contractor's performance during stage two.
- The Client's discretion on whether to proceed to stage two at the end of stage one was absolute.
- The Client also had the right to terminate the contract
  - if the Contractor failed to submit proposals for stage two which met the gateway review criteria
  - at any other point during stage one, at its absolute discretion.

Much of the negotiation between the parties focused on the amount of compensation payable by either party in each of these termination scenarios.

- The Client was entitled to amend the contract at any time during stage one in order to include additional incentive payments during stage two and/or to introduce, during stage two, an alliance agreement with other key package Contractors, provided this did not expose the Contractor to material additional cost, liability or performance risk.

4.3.5   Target price pain/gain share incentive schemes

The NEC4 Option C model

The NEC drafting on the target price pain/gain share regime is set out within Option C clause 54. (If the Contract also includes Option X22, allowing the parties to use stage one to arrive at a properly interrogated Target Price for stage two, then it would apply in relation to that stage two Target Price.) The drafting is very clear and, as such, it does not usually attract much debate between the parties. If the Price for Work Done to Date (which is defined as the amounts payable to the Contractor on account of its Defined Costs and Fee entitlement) is less than the target Prices (which are defined as the lump sum prices for each of the activities on the Activity Schedule), the Contractor is paid its share of the saving; if they are greater than the target Prices, the Contractor pays its share of the excess. Details of the share percentages and ranges are set out in the Contract Data. A preliminary assessment of the Contractor's share is made on completion of the works and included in the payment due to the Contractor after completion. A final assessment is made and paid once the final Price for Work Done to Date has been ascertained.

Contract Data, Part One

Assume the target Prices are £100 000.

Assume share range and percentages are as follows in Table 4.1.

Tables 4.2 and 4.3 set out two worked examples, showing how the share percentages and ranges might operate.

Table 4.1. Share range and Contractor's share percentage assumptions

Share range	Contractor's share percentage
less than 80%	15%
from 80% to 90%	30%
from 90% to 110%	50%
greater than 110%	20%

Example 1

Final Price for Work Done to Date (i.e. Defined Cost plus Fee) = £75 000

Saving therefore = £25 000

Contractor's gain share to be paid by Employer is £8750, calculated as shown in Table 4.2.

Example 2

Final Price for Work Done to Date (i.e. Defined Cost plus Fee) = £115 000

Cost overrun therefore = £15 000

Contractor's pain share to be paid or allowed to the Employer is £6000, calculated as shown in Table 4.3.

Table 4.2 Contractor's gain share calculations

Share range	Amount of saving	% .... = £
less than 80%	£5000	15% = £750
from 80% to 90%	£10 000	30% = £3000
from 90% to 110%	£10 000	50% = £5000

Table 4.3 Contractor's pain share calculations

Share range	Amount of saving	% .... = £
less than 80%	Not applicable	
from 80% to 90%	Not applicable	
from 90% to 110%	£10 000	50% = £5000
greater than 100%	£5000	20% = £1000

Case studies

All the debate between the parties concerning the pain/gain share incentive arrangements in our case studies focused on one or more of the following related commercial issues

- those cost items which fall within the definitions of 'Defined Costs' and 'Disallowed Costs' and the extent to which the Client should have control and/or oversight of the Contractor's supply chain arrangements

- whether, for the purposes of calculating the Contractor's pain or gain share
  - o the estimate of the Contractor's total Fee entitlement should be included within the target Prices set out within the Activity Schedule, given that the Contractor's actual Fee entitlement is included within the Price for Work Done to Date
  - o any account should be taken of any KPI deductions that are made from the Price for Work Done to Date
- whether the pain share payment provisions should be amended so that, once the Price for Work Done to Date has exceeded the target Prices (or perhaps, when the Project Manager concludes that there is no reasonable prospect of the Price for Work Done to Date being less than the Prices), the Project Manager is able to deduct from all further payment certificates amounts equal to its estimate of the Contractor's ultimate pain share
- the extent to which Defined Cost payable in respect of compensation events should increase the target Prices and thereby preserve the incentive
- whether any compensation events which reduce the Contractor's Defined Costs should also reduce the target Prices (other than a change to the Scope to give effect to a value engineering proposal submitted by the Contractor, which is provided for in Option C clause 63.13)
- the extent to which recoveries made by either party under project insurances taken out by the Client should be taken into account for the purposes of the pain/gain share calculation
- the percentages at which and the bands within which any pain and gain is shared between the parties
- any circumstances in which the Fee should be no longer payable on the Contractor's Defined Costs
- the extent to which any Contractor exposure to pain share counts towards any limit on its aggregate liability under the contract.

4.3.6 Alliance agreements and project-wide incentive schemes

In our experience, while project-wide alliance arrangements and attendant incentive schemes are used on some of the most complex, disaggregated engineering projects, they are not currently in widespread use within the UK construction industry. One can speculate as to the reasons for this. It may be that project sponsors and their lenders question whether such schemes actually deliver value for money, based as they are on initial capital cost savings rather than whole life value across the lifetime of a project. Or it may simply be that they believe that it is only the biggest, disaggregated projects, based on a cost plus contracting model, that justify the time and expense involved in negotiating and concluding and then administering such alliancing arrangements.

The NEC4 Option X12 model (Multiparty Collaboration)

Option X12 is designed to construct a partnering arrangement between the Client, the Contractor and those other project participants who are able to influence the achievement of the

Client's objectives (as stated in the Contract Data) and the Partners' objectives (as stated in a Schedule of Partners). A Core Group of the Partners (specified in a Schedule of Core Group Members) acts and takes decisions on behalf of the Partners in relation to those matters that are stated in the Partnering Information. The Contractor gives effect to the decisions of the Core Group and any consequent change to the Partnering Information and/or the Contractor's programme will be a compensation event and may therefore increase or reduce the Prices.

The Contractor is obliged to work together with the other Partners in a spirit of mutual trust and cooperation to achieve the stated objectives, to provide information to the other Partners, to give early warnings to the other Partners if it becomes aware of any matter likely to affect the achievement of any of the objectives and to use compatible information management systems.

KPIs and the corresponding performance targets are stated in the Schedule of Partners. If the target for a KPI is met or exceeded the Contractor is paid the amount stated in the Schedule of Partners.

However, there are two important points to recognise when using the Option X12 model.
- It does not create a multiparty contract – there is no legal relationship between the Contractor and the other Partners and there is no Client obligation to procure performance by the other Partners of their corresponding obligations to the Client, so the effective operation of an Option X12 alliancing arrangement will depend on the existence of a continuing consensus between all partners.
- It may be difficult to maintain that consensus if there is any conflict between the Works Information and the Partnering Information – an objective which may be difficult to achieve if (as seems likely) the Client does not know how the Alliance will actually operate at the point in time when it has to settle the content of those two important documents.
- Each Partner contract will need to be concluded on NEC4 terms that are the same in all material respects and which have been amended as appropriate in order to give the Alliance the flexibility it needs to operate effectively.

The NEC4 Alliance Contract model
This point was not lost on the NEC when it published its Alliance Contract in June 2017. The contract recognises that, if an alliance is to operate as intended, it has to recognise three fundamental truths.
- A successful incentive regime depends on a successful Alliance.
- An Alliance can only exist where there is consensus.
- The Alliance parties can only be incentivised to achieve an outcome if they are able to influence that outcome by their performance.

In essence, the Alliance Contract adopts the NEC4 Engineering and Construction Contract Option X12 to form the core of a single, multiparty agreement. The main components of the Alliance can be summarised as shown in Figure 4.1.

Figure 4.1 Principal components of the Alliance (author's own)

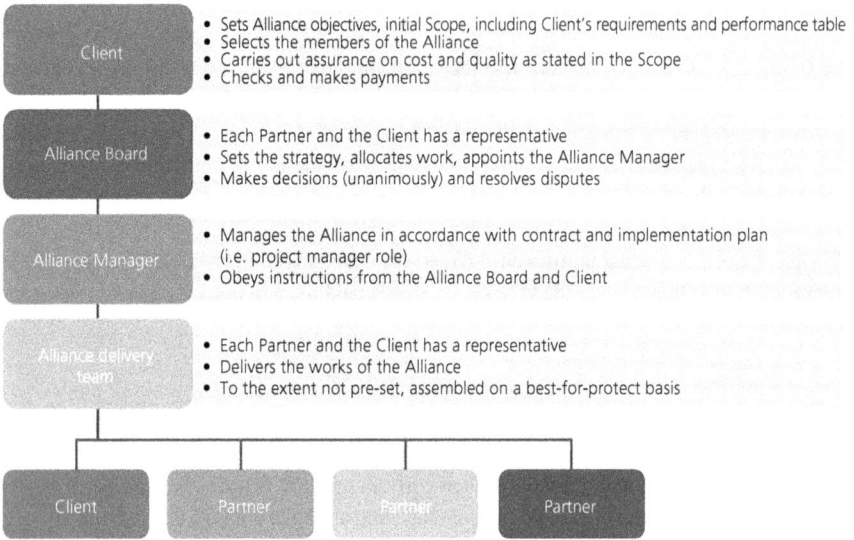

Client
- Sets Alliance objectives, initial Scope, including Client's requirements and performance table
- Selects the members of the Alliance
- Carries out assurance on cost and quality as stated in the Scope
- Checks and makes payments

Alliance Board
- Each Partner and the Client has a representative
- Sets the strategy, allocates work, appoints the Alliance Manager
- Makes decisions (unanimously) and resolves disputes

Alliance Manager
- Manages the Alliance in accordance with contract and implementation plan (i.e. project manager role)
- Obeys instructions from the Alliance Board and Client

Alliance delivery team
- Each Partner and the Client has a representative
- Delivers the works of the Alliance
- To the extent not pre-set, assembled on a best-for-protect basis

Client | Partner | Partner | Partner

Other points to note are as follows.
- It largely follows the structure of the NEC4 Engineering and Construction Contract.
- It is a performance-based contract – there is risk and reward for achieving or failing to achieve the Alliance Objectives set by the Client (clause 53).
- There is an early Alliance involvement Option X22.
- All members of the Alliance Board have an equal voice.
- It is a cost reimbursable contract – Defined Costs plus a Fee are assessed by the Alliance Manager, certified by the Client's Representative and paid into a Project Bank Account.
- Compensation events are limited to Client or Alliance Board changes to the Client's Requirements (Client information comprising part of the Scope), denial of site access, instructions to suspend and force majeure events. Note, in particular, that Alliance changes to the remainder of the Scope are not compensation events (unless required as a result of a compensation event).
- Adjustments to the Budget, the Completion Date and the Performance Table are made in respect of all compensation events.

- All liabilities are shared by the Alliance, save for
  - willful default by a Partner
  - breach of intellectual property rights
  - third-party liability arising from work carried out by a Partner
  - liability for death of or personal injury to a Partner's employees.

Third-party claims made against the Client in respect of work carried out by members of the Alliance are borne by the Client and are not the subject of an indemnity from the relevant Partner(s), unless they arose as a result of work carried out by a Partner in willful default of its obligations. However, to the extent that the project insurances do not respond to cover any liability arising in respect of such claims, it will be included within the Alliance Cost and will therefore adversely affect the Alliance's pain or gain share.

As one would expect with an Alliance agreement, all disputes are resolved by the Alliance Board in an endeavour to promote a 'no blame, no claim' culture. Failure to accept the Board's decision will result in the removal of the disaffected Partner from the Alliance (if all other Partners agree) or termination of the contract (if all Partners agree).

The Alliance Board may terminate a Partner's participation in the Alliance if it is in default and all other Partners agree. It may also terminate the contract if
- the Client is in certain specified default
- performance is frustrated by force majeure or
- all Partners agree to terminate and the basis on which the termination will take effect.

If used, the Alliance Contract will replace the Client's contracts with the individual Partners, thereby replacing the individual pain/gain share incentive regimes within those contracts with an overarching, project-wide incentive scheme linked to project objectives and performance.

This is an important consideration. As previously stated, the continued success of the Alliance will depend on a continued consensus between the Partners. In the absence of that consensus the contract will be terminated, leaving the Client to restructure the project. The nature of that risk may lead a Client to question whether it is worth spending the time and money necessary to conclude an Alliance Contract between several different package Contractors who may never have been party to such an agreement before. Conversely, an alternative, two-tier incentivisation regime based on NEC4 Option C and Option Clause X12 would leave the Option C pain/gain share incentive in place, even if the Option Clause X12 Alliance incentive breaks down.

We believe that an Alliance arrangement is only really justifiable for very large and complex projects, where the programme, design solution and consequent work-scope requirements cannot be settled with a degree of certainty that will deliver pricing which represents value for money. In our experience, the difficulty inherent in attempting to persuade a supply chain to engage in a process which involves considerable, up-front investment of resource and the sharing of risks and rewards means that it is only those projects with the biggest carrots that will stand a reasonable chance of successfully concluding such arrangements.

Case study 1

On this project, which involved the procurement of a major infrastructure asset within the energy sector on a construction management basis, the Client entered into several major works package contracts, all of which were based on the NEC3 Engineering and Construction Contract, Option C. The Client, each of the package Contractors and the Construction Manager appointed by the Client also entered into a bespoke Alliance Agreement, pursuant to which

- all participants committed to work collaboratively, to establish an integrated team environment and to delegate decision making in all matters relating to the interface and integration of the package contract works to an Alliance Board which would operate on a 'best for project' basis – that is, on a basis consistent with the achievement of the Alliance's stated objectives to
  - o achieve project milestones and completion of the whole project by the planned completion dates
  - o complete the project within the forecast budget
  - o undertake efficient procurement and deployment of shared equipment, materials and facilities
  - o mitigate project delays and cost overruns
  - o maximise early contractor involvement in an initial design development process
- the Construction Manager would implement and the Contractors would comply with all unanimous decisions of the Alliance Board without any compensation or relief entitlement under the works package contracts (unless otherwise directed by the Alliance Board)
- the Contractors and the Construction Manager would share in any saving made against a target cost for the Project, receive a bonus payment if the Project was completed ahead of a planned completion date and receive a quarterly KPI bonus payment if certain performance targets were met or exceeded
- none of them would be exposed to any target cost pain or project programme over-run
- there would be no adjustment to the project target cost and/or the planned project completion date (unless the Client, at its absolute discretion, decided to do so)
- each of the Contractors would, if so instructed by the Client, step in and complete any other Contractor's works, in the event of that Contractor's failure to do so.

The parties also agreed that, save in relation to direct loss arising as a result of fraud, death or personal injury caused by a party, any breach of the Client's obligation to make incentive payments or any failure by a Contractor to implement a unanimous board decision or to complete another Contractor's works (if instructed to do so by the Client), they would have no liability to each other under the Agreement.

The Alliance Agreement was also supported by a cooperation agreement, setting out the procedures which the parties would operate in order to achieve the Alliance objectives and preserve the Alliance incentives.

Case study 2

On this project, which involved the procurement of a major infrastructure asset within the transport sector, also on a construction management basis, the Client wished to adopt a procurement structure which would recognise the complexities inherent in a disaggregated, multicontract procurement involving multiple systems interfaces and accommodate the lessons learnt on similar previous projects by allocating risk to those who are best able to manage that risk.

This approach led to a procurement structure based on a split work scope.

- Risk in the cost of 'package scope work' was placed with the Contractors, by way of the operation of core contracts based on the NEC3 Engineering and Construction Contract, adopting an Option C target price, pain/gain share arrangement
- Risk in the cost of 'alliance scope works' was ultimately retained by the Client, although controlled by way of the operation of an Alliance Agreement (populated by the Contractors), pursuant to which the Alliance was paid at cost with joint, upside only, incentives to deliver to cost and programme.

The 'alliance scope works' primarily consisted of the common, project-wide, systems installation, testing and commissioning elements of the works. In relation to those works, the Client sought to construct an Alliance Agreement based on the answers to the following key building blocks.

- What is the Alliance being asked to do?
- How will the Alliance be operated and established?
- What information is to be provided to and generated by the Alliance?
- How will the Alliance be governed and how will it make decisions?
- How will the Alliance be incentivised or rewarded?
- How is the performance of the Alliance members to be addressed?
- How will the NEC3 EEC contracts need to be adapted in order to accommodate the operation of the Alliance?

The contract structure for the Alliance and its key features can be summarised as shown in Figure 4.2.

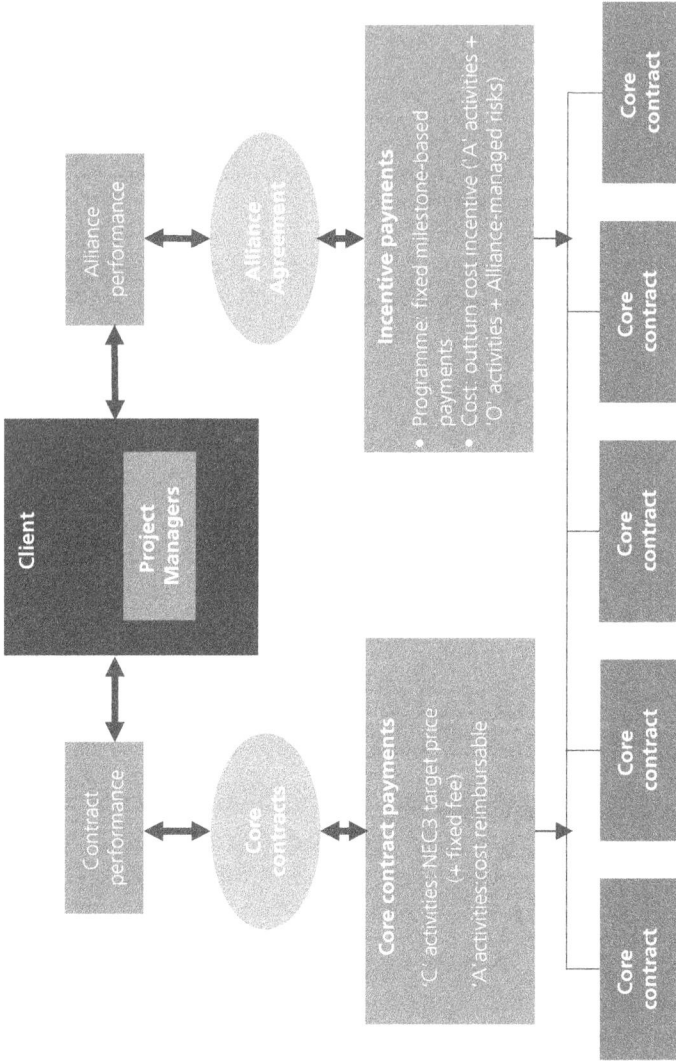

Figure 4.2 Alliance structure and key components (author's own)

Definitions

- 'A' activities are the Alliance scope works
- 'C' activities are the package scope works
- 'O' activities are the operational (i.e. people) activities of the Alliance

Core contract key features

- hybrid NEC3 Option C Target Price, amended as necessary to interface with Alliance Agreement
- programme managed at the Alliance level
- fixed fee
- KPIs for Contractor performance, with consequent fee adjustments.

Alliance Agreement key features

- role: delegated authority on 'A' activities and advisory on 'C' activities
- resourced by Contractor and Client personnel, selected on a 'best athlete' basis
- single point of formal communication and administration control by way of the core contract Project Managers
- Client responsible for managing external interfaces with the Alliance, including risk of consequent scope and programme changes.

Incentive arrangement key features

- The achievement of major integration milestones is incentivised, with the Contractors taking an equal share of the incentive amount on the achievement of each milestone.
- The Alliance budget includes (*a*) the target cost of 'A' activities, (*b*) an allowance for 'O' costs and (*c*) a contingency allowance for Alliance-managed risks.
- The contingency allowance within the Alliance budget is used to fund compensation events under the core contracts that arise as a result of an Alliance-managed risk.
- An agreed percentage of the Client's share in any gain against the Target Cost under a core contract is added to the contingency allowance for Alliance-managed risks.
- Net savings achieved across the three elements of the Alliance Budget are shared in an agreed percentage between the Client and the Alliance. The Alliance's share is split equally between the Contractors and paid on completion of the final incentivised programme milestone.

REFERENCES

FIDIC (Fédération Internationale des Ingénieurs-Conseils / International Federation of Consulting Engineers) (2017) *Red, Yellow and Silver Books*. FIDIC, Geneva, Switzerland. https://fidic.org/themes/new-fidic-contracts-2017-2nd-editions-red-yellow-and-silver-books (accessed 09/07/2024).

JCT (Joint Contracts Tribunal) (2016) *Contract Families*. JCT, London, UK. https://www.jctltd.co.uk/category/contract-families (accessed 090/07/2024).

NEC (New Engineering Contract) (2016) *How to provide early contractor involvement using an ECC*. NEC, London, UK.

NEC (2017) *NEC Contracts*. NEC, London, UK. https://www.neccontract.com/ (accessed 09/07/2024).

emerald PUBLISHING   ice

Michael Smith, Matthew Finn and Jon Broome
ISBN 978-1-83549-897-2
https://doi.org/10.1108/978-1-83549-894-120241005

Chapter 5
Realising the practical benefits

Matthew Finn

Abstract

This chapter
- describes the benefits to enhance the design and maximise collaboration while developing and defining the scope of works
- describes the benefits to the delivery of the project and programme
- considers the collaboration of the team to ensure shared incentives linked to the success of the project and key deliverables
- considers the improvement in risk management of the parties through shared ownership and early identification of risks
- explores subcontract considerations when using early contractor involvement (ECI)
- explores the use of building information modelling (BIM) with ECI
- identifies case study examples of the successful application of ECI.

5.1.   Introduction

In this chapter, we focus on the practical benefits of early contractor involvement (ECI). ECI is intended to benefit a project in three major ways: tighter timescales, reduced cost and better quality. The uptake of ECI within the UK construction industry, particularly on large, complex projects, has been steady. However, recommended methodology and best practice remains a grey area.

It is generally recognised that the ability of the parties to influence project outcomes, including reduction of cost, creation of additional value, improvement of performance and flexibility to incorporate changes is much higher in the earlier conceptual and design stages of the project (Mosey, 2009).

A variation of the ECI model has been adopted widely in the UK infrastructure sector on large and complex projects, namely optimised contractor involvement (OCI). The subtle difference between ECI and OCI is that in ECI, the contractor's skills are introduced early into a project to bring design buildability and cost efficiencies to the preconstruction phase. In OCI, contractors are brought in at a stage that is late enough for the target price to be quite firm, but early enough for them to be able to influence buildability and value engineering (Designing Buildings, 2020).

5.2. Benefits

5.2.1 Introduction

ECI gives the contractor a point of responsibility early in the project life cycle which the client therefore benefits from, enhancing the design and maximising collaboration while developing and defining the scope of works, leading to overall better quality through buildability issues being addressed early. This early involvement assists in achieving cost-effective, high-quality outcomes while minimising risks and delays, when such risks are endemic in mega projects, such as the Channel Tunnel, Boston's Big Dig, Sydney Opera House and so on.

A key driver in the ECI approach is the collaboration of the team to ensure shared incentives linked to the success of the project and key deliverables. Mosey notes that

> …commentators have recognised that a procurement model which omits contractor and specialist design contributions can increase risk and can result in poor communications between team members, unnecessary delays to the progress of the project and the creation of incorrect information that leads to claims and disputes (Mosey, 2009: p.7).

5.2.2 Programme for delivery

ECI as a project management approach offers benefits to the delivery of the project, such as health and safety improvements, ability to accelerate and contractor input into the programme to provide a more realistic and detailed programme for delivery.

Delay in the creation of a programme until after start on site is widely accepted in the construction industry, even although such delay means that the contractor at the point of start on site is immediately under pressure to create an additional document that is necessary for it to work in an efficient manner, with the risk that adoption of this document will be further delayed if it is not accepted by the client or project manager or the contractor's own supply chain (Mosey, 2009).

It is difficult to enforce a plan which is conceived in isolation, and it is, therefore, essential to involve the individuals and organisations responsible for the activities or operations as the plan is developed. However, programming is often left too late. Time pressures on many projects mean that the set of coordinated method statements, programmes and budgets will still be under development after construction has started (See Mosey (2009) for further discussion).

Risks that arise if contractors do not have an involvement in the earlier activities include: programme – the construction phase programme is not agreed on until after start on site, creating uncertainty as to key dates required for activities, such as the release of further consultant or contractor design details, and for the pricing and approval of provisional sum items. With ECI, the construction phase programme can be agreed prior to starting on site, including key dates for activities, such as release of remaining consultant and contractor design details and the pricing and approval of provisional sum items.

Unrealistic time targets can cause disputes and are directly linked to activities undertaken during the preconstruction phase of a project; the risk of these causes arising can be avoided or reduced by collaborative activities undertaken through ECI. Unrealistic time targets can be challenged if team members have an opportunity for joint programming during the preconstruction phase, agreeing to deadlines and interfaces in respect of construction phase dates rather than leaving these to be established by unilateral client or consultant decisions.

A key function of a programme governing the preconstruction phase will be the completion of all activities that are preconditions to proceeding with the construction phase, including the finalisation of agreed designs, prices and supply chain arrangements, the establishment of an acceptable understanding regarding project risks and a variety of other matters, such as satisfying health and safety requirements, obtaining third-party consents, securing full project funding and agreeing the construction phase programme.

5.2.3   Buildability

CIRIA (1983) carried out a preliminary study into 'buildability' and noted that the 'achievement of good buildability depends upon both designers and builders being able to see the whole construction process through each other's eyes' (CIRIA, 1983: p.3). The problem with buildability was probably due to 'the comparative isolation of many designers from the practical construction process' (CIRIA, 1983: p.6). A definition of buildability was agreed for the purposes of the study as 'the extent to which the design of a building facilitates ease of construction, subject to the overall requirements for the completed building' (CIRIA, 1983: p.6).

Good buildability is a direct consequence of design decisions. On that basis, buildability insights need to be introduced into the design at an early stage. ECI enables a main contractor and important subcontractors to join the project team at an early stage to offer a buildability view within the overall development of the design (CIRIA, 1983). CIRIA recognised that

> good buildability represents an overall financial benefit to the client. In addition, it leads to lower costs for the builder because work is more systematic, and it may also lead to lower costs for the designer for the same reason (CIRIA, 1983: p.7).

Construction projects are increasingly highly mechanised and complicated, and even the very best design consultants cannot have all the detailed knowledge available to them that is available to specialist contractors through their research and development departments and through their on-site project teams.

The case for main contractor and subcontractor participation in design development is founded on the proposition that, in many cases, design consultants alone cannot develop designs that

- are sufficiently detailed to be capable of comprehensive fixed pricing by a main contractor in a single stage tender
- are fully buildable by a main contractor without further detailing and/or amendment to reflect circumstances on site and the interaction between various trades
- embody the latest thinking of manufacturers, suppliers and specialist trades (Mosey, 2009).

The JCT Constructing Excellence Guide states

It is important that contractors and any key specialists are engaged early, ideally at a stage when the proposed design is not complete so that it is possible for the contractor and key specialists to consider ways in which the design can be made easier to build and maintain' (JCT, 2016: section 38:7).

Banwell (1964: p.4, section 2.6) recognised the need to appoint the main contractor 'before … the programme of work [is] finally settled' and suggested that there is a role for the main contractor alongside the rest of the team in completing missing details before embarking on the construction phase of the project.

Banwell (1964) suggested that, in many cases, design consultants cannot be confident that their designs adopt the most practical and economical approach without the opportunity to review those designs in conjunction with the organisation or organisations that are going to implement them on site. For this to be done properly, and for cost and time benefits to be achieved, Banwell recognised the need for a significant period prior to the commencement of the construction phase, during which the main contractor is appointed to the team. While Banwell's recommendation of main contractor contributions to design might be interpreted simplistically as a case for 'design and build', his proposal that the client should appoint the main contractor to assist in finally settling the programme of work is a clear argument for joint working during the preconstruction phase (Mosey, 2009).

5.2.4   Appointing subcontractors

There needs to be careful consideration surrounding the engagement of subcontractors by the contractor under an ECI, such as when to appoint subcontractors and on what basis, both in terms of selection and contract arrangements – for example, consultancy contract initially, lump sum or target cost basis and who decides the basis. The subcontractors, in particular specialist suppliers, are vital to the contractor's coordination and management of the ECI; therefore, managing and utilising their subcontractor's expertise during the contract has to form part of the procurement process and programme for the project.

Joint client/main contractor selection of subcontractors and suppliers marks a departure from the traditional assumption that a subcontractor or supplier is either 'domestic' when selected by the main contractor or 'nominated' when selected by the client. Joint selection of subcontractors and suppliers as part of a structured preconstruction phase process avoids the need for nomination and thereby also avoids consequent confusion as to the extent of the client's and the main contractor's liability for such subcontractors and suppliers. It permits direct client and consultant influence over which subcontractors and suppliers are most appropriate and offers the best value, combined with acceptance of the main contractor's responsibility for their subsequent performance during the construction phase. Any sharing of risk for subcontractor or supplier default or insolvency can be agreed between the client and main contractor as part of the preconstruction phase process or later in response to specific circumstances (see Mosey (2009) for further discussion).

A problem in the early selection of subcontractors and suppliers as design team members is how best to obtain their design and risk management contributions while retaining a competitive process for their appointment to implement the works package they have designed. The options that emerge will be more or less appropriate according to the nature of the works package, the length of the preconstruction phase and the commercial preference of the clients, consultants and main contractor.

A further question for the client and main contractor to address is: what are the most appropriate criteria that should govern the selection of subcontractors and suppliers? Aside from the price components that the client is seeking to fix, the criteria for the selection of subcontractors and suppliers should be consistent with those used for the selection of main contractors. Otherwise, the temptation of clients and main contractors to focus primarily on subcontractor or supplier prices over and above other considerations may lead to an imbalanced team where the selection of subcontractor and supplier members has not reflected the client's wider project priorities.

Finally, the client and main contractor need to agree on whether subcontractors and suppliers should be paid for their preconstruction phase contributions to the project. For the client to obtain the greatest value from the design and other contributions of subcontractors and suppliers, a commercial incentive of some kind is likely to be necessary.

Risks that arise if contractors do not have an involvement in the earlier activities include: for subcontract tenders – main contractor bidders need to require their respective subcontractor bidders to assimilate complex project information within a limited time and with a proportionally remote prospect of success, and to propose complete and binding subcontract prices; for subcontractor appointments – subcontractor appointments are likely to be finalised only after start on site, allowing the potential for repricing and change of subcontractors (with no benefit to the client) if the main contractor seeks to increase its profit by finding cheaper subcontract deals or if subcontractors withdraw before they are signed up.

5.3.   Risk management

The key benefit of ECI is the opportunity to have an earlier and better risk management of the parties through shared ownership and early identification of risks. Contractors and sub-contractors bring practical experience to the table and can identify potential risks associated with construction activities. By addressing risks early in the project, it is possible to develop mitigation strategies and avoid delays or unexpected issues later on.

With traditional delivery methods, it is usually consultants that will undertake project risk assessments on behalf of the client. The assumption then is that, prior to submitting its tender, each bidding contractor will undertake an assessment of risk at its own cost and risk. The bidders are left to interpret the client's and consultant's project risk assessments when formulating their bid prices for the project. This approach does not enable the bidding contractor or its proposed supply chain to interrogate the client's and consultant's risk assessment. It also does not allow the client, its consultants and the selected contractor and its proposed supply chain time to work together to reduce risks and the prices attached to those risks (Mosey, 2019).

A client and its consultants have considerable time to assess risks, whereas a bidding contractor has only a period of a few weeks to undertake its separate assessment while at the same time compiling all other aspects of its response to the client's invitation to tender (see Mosey (2019) for further discussion).

In a single-stage competitive approach, bidding contractors are unlikely to fully understand all aspects of the proposed project or have sufficient time to identify and consider how to manage potential risks to the project before submitting its tender. The tenderers' risk assessments are likely to be abbreviated and influenced by their wish to submit successful responses to the technical and pricing requirements of the bid.

The client and its consultants may decide that any bidder proposals describing risk assessments that differ from those in the invitation to tender are in fact techniques to gain a more favourable financial position. Also, each bidding contractor has no right or responsibility to analyse risk jointly with the client and consultants, and this can lead to
- a cynical assessment by bidders of client and consultant risk assumptions
- the exploitation by the selected contractor of weaknesses in the client's documents
- the exploitation by the selected contractor of any risks which the client has not comprehensively transferred (Mosey, 2019).

For any team to manage risk effectively, it needs to establish the time and processes necessary to undertake joint risk management exercises after contractor selection and prior to start on site, and the ability to make adjustments that reflect the outcome of these exercises.

If procurement models, contracts and consultant risk appraisals focus only on the transfer of risk and not on its management, this usually gives rise to a risk premium or contingency charged by the party accepting the transferred risk. In addition

> the fact that both supply chain risk allowances and client contingencies are calculated more by rules of thumb than by a reliable scientific means creates the inherent problem that risks may not be adequately covered by either individual risk allowances or client contingencies (JCT, 2016: section 33).

A commercial issue that needs to be tackled is the fact that, as additional information is built up following an early contractor appointment, it will not be possible for the client to transfer risks that emerge later in the preconstruction phase of the project if the contractor is not willing to accept them (Mosey, 2009). Risk management is not an orderly sequential process comparable to other preconstruction activities and it is not possible to guarantee in advance that joint risk management involving the contractor will lead to a risk and cost position acceptable to all parties. However, for the client and its advisors to seek fixed prices from a main contractor without recognising the scope for it to contribute to early risk management is to draw a veil over important commercial factors. Specifically, under a traditional single-stage contractor appointment, if risks arise during construction which the main contractor has not foreseen at the time of its tender or if a risk contingency allowed by the main contractor proves to be insufficient, it is unlikely that the main contractor will allow a profitable job to become loss-making simply because it accepted those risks within its fixed price. Instead, this situation is likely to give rise to manoeuvring and claims by the main contractor to try to recoup any loss deriving from its miscalculation. This in turn can be prejudicial to the quality of the project – for example, if the main contractor looks for ways of cutting costs that may not be in the interests of the client and may not be declared to the client (Mosey, 2009).

Where contracts continue to focus only on the transfer of risk and not on its management, this will usually give rise to a risk premium charged by the party accepting the transferred risk. The risk premiums or contingencies allowed by main contractors and their supply chain members are likely to be broad estimates, and can prove insufficient to cover the cost of the required remedial actions if and when the risks materialise. However, in those circumstances, it is likely that any contractor, subcontractor or supplier will be unwilling to incur the additional costs necessary to cover the risks as this will erode its profit. As a result, the project will suffer from the claims and counter-claims that arise from the client seeking to impose risk transfer provisions and from the contractor, subcontractors and suppliers resisting costs that make the project unprofitable. In these circumstances, the client and the project are likely to suffer adverse consequences far greater than the cost of the client retaining the risk or agreeing a joint strategy with the contractor and other team members for managing that risk (Mosey, 2019).

Successful risk management involves allocation of risks to the contractor that it is able to manage them. Such risks should not be allocated on the basis of expediency, which can be the result of a priced-based single-stage tender. In the long run, it is better value for a client to pay for risks that actually occur during the construction phase of a project rather than to agree a price based on what a contractor thinks might occur. In the latter case, risk is transferred arbitrarily and both the client and the contractor are gambling on whether that risk has been accurately costed (see Mosey (2019) for further discussion).

Where unanticipated risks have a later impact on the expected designs or methods of working, the parties are left to argue as to who bears the cost and time consequences and whether the contractor has a claim by reason of an express contract provision in relation to the risk in question or by reason of information provided by the client being incorrect or incomplete.

Risks that arise if contractors do not have an involvement in the earlier activities include: for joint activities – any post-tender reviews by the selected contractor of the client's and consultant's design and risk assumptions will delay start on site, allowing very limited scope for any joint client, consultant and contractor activities, such as value engineering or risk management to reduce excessive costs or to resolve consultant design errors revealed in main contract tenders. Such reviews also become confused with commercial negotiation of the contractor's price (Mosey, 2009).

Risks that arise if contractors do not have an involvement in the earlier activities include: for risks – risks are assessed by the client and the consultants, making assumptions as to the likely interpretation of such risk assessments by the main contractor and its subcontractors (Mosey, 2019).

There is a perception that significant recent events in the UK (for example, the demise of Carillion and the difficulties which led Interserve to exit the energy from waste sector) have demonstrated the potentially disastrous consequences of a single entity taking the entire project delivery risk without properly understanding the nature of that risk and making appropriate allowance for it. Stepping down the whole of the risk to a single contractor who has priced that risk in a highly competitive environment may no longer be seen as representing value for money (and may be considered unwise in certain technologically challenging sectors). A two-stage procurement model has been used and refined by those involved in the structuring of some of the largest and most complex infrastructure projects within the UK over recent years, where there was no market appetite for a single, turnkey entity – for example, Crossrail, Thames Tideway, Hinkley Point C and High Speed 2 (Smith, 2019).

5.4. Building information modelling (BIM)

Through ECI, contractors may bring innovative construction methods or technologies to the project, leading to improved efficiency, sustainability or safety. The use of building information modelling (BIM) provides the opportunity to integrate these innovations seamlessly into the project plan. The integration of BIM with ECI enhances collaboration, improves decision making and contributes to the overall efficiency of construction projects. It allows for a more informed, collaborative and streamlined approach to the design, construction and management of built assets. There is an opportunity at an early stage of a project to undertake risk management through clash detection and better coordination as well as giving three-dimensional (3D) visualisations, making it easier for all parties involved to understand project details and make informed decisions, thereby reducing the probability of later changes/variations to the project resulting in over-run and overspend of the project.

The 2011 UK Government Construction Strategy (Cabinet Office, 2011) set out to introduce a progressive programme of mandated use of fully collaborative BIM for centrally procured government construction projects by 2016. BIM Level 2 was developed to meet this mandate (Cabinet Office, 2016). The distinguishing aspect of this level was collaborative working with streamlined information exchange and coordination between all stakeholders and systems.

BIM was identified in the 2016 UK Government Construction Strategy as a

collaborative way of working that facilitates early contractor involvement, underpinned by the digital technologies which unlock more efficient methods of designing, creating and maintaining our assets (Cabinet Office, 2016: section 22).

Within the engineering and construction arenas, new standards, innovations and technologies are continually transforming BIM. In 2019, the UK BIM Framework (2019) was launched, replacing some of the existing British Standards and Publicly Available Specifications. The UK BIM Framework is based on international standard BS EN ISO 19650 (BSI, 2019) which was developed out of the UK's BIM Level 2, but also embraces international digital perspectives. Therefore, to align with the ISO 19650 series, the use of notations such as 'BIM Level 2' are being discouraged, in favour of the 'UK BIM Framework'.

BIM can go further and be used for Level 4 (Time) and Level 5 (Cost); therefore, combining BIM with scheduling information and cost data provides a visualisation of the construction timeline and associated costs. The benefit for the contractor is that this gives a better understanding of project sequencing, resource allocation and cost implications, contributing to more informed decision making for its resources, preliminaries and choice of supply chain.

In May 2020, a new Information Protocol was published by the UK BIM Framework (2020) to support contracts which use ISO 19650. Previously known as a BIM Protocol, the Information Protocol is a document required under ISO 19650 that agrees the various roles, responsibilities and rights of the parties.

Most standard form contracts make only light reference to BIM within their documentation. NEC4 arguably goes the furthest. The Information Protocol can be incorporated into the

NEC4 ECC by the selection of secondary Option X10 and the inclusion of the relevant parts of the Protocol as the Information Model Requirements in the Scope (NEC, 2020).

5.5. Case studies

Two-stage procurement model

Thames Tideway

Tideway chief technical officer Roger Bailey believes that the undoubtedly successful construction of London's super sewer – the 25 km-long Thames Tideway Tunnel – is largely down to the upfront collaborative decisions taken before construction began in 2016 (Parker, 2022).

SCAPE framework

The SCAPE Group is a public-sector organisation which attracts some of the UK's largest contractors to deliver SCAPE's 'combined buying power of £13 bn, our frameworks have delivered over 2400 successful projects' (SCAPE, 2020).

SCAPE has provided the author with data on its projects which have used the NEC ECI X22 as part of its procurement, of which the values are set out in Figure 5.1.

Figure 5.1 Use of NEC ECI X22 on the SCAPE framework (author's own)

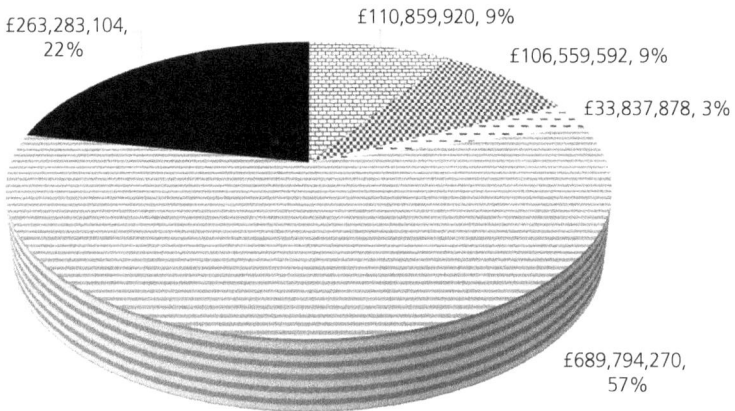

£263,283,104, 22%

£110,859,920, 9%

£106,559,592, 9%

£33,837,878, 3%

£689,794,270, 57%

⋈ Major Works England, Wales and NI 2017 contracted to Wilmott Dixon Construction Limited
■ Major Works UK contracted to Wates and McLaughlin & Harvey
⠿ Major Works Scotland contracted to Roberson
≡ National Civil Engineering and infrastructure (Gen 1) contracted to Balfour Beatty
■ National Civil Engineering and infrastructure (Gen 2) contracted to Balfour Beatty

As shown in Figure 5.1, circa £953 m, the vast majority of the £1.2 billion spent has been on the civil engineering frameworks with Balfour Beatty. When Balfour Beatty was appointed to the renewed framework in 2022 it commented

Utilising early contractor engagement, Balfour Beatty will ensure that the best value design solutions are in place, driving efficiencies and helping to transform local communities and support both regional and national economic growth (Kanaris, 2022).

Therefore, a large amount of the success of the SCAPE framework seems to be attributed to the ECI model in its procurement.

REFERENCES

Banwell CH (1964) *The Banwell Report: The Placing and Management of Contracts for Building and Civil Engineering Works*. Her Majesty's Stationery Office, London, UK.

BSI (2019) BS EN ISO 19650: Managing Information with Building Information Modelling (BIM). BSI, London, UK.

Cabinet Office (2011) Government Construction Strategy. Cabinet Office, London, UK.

Cabinet Office (2016) Government Construction Strategy 2016–20. Cabinet Office, London, UK.

CIRIA (Construction Industry Research and Information Association) (1983) *Buildability: An Assessment*. CIRIA, London, UK, Special Publication 26. https://www.ciria.org/CIRIA/CIRIA/Item_Detail.aspx?iProductCode=SP26D&Category=DOWNLOAD (accessed 11/07/2024).

Designing Buildings (2020) *Optimised Contractor Involvement*. https://www.designingbuildings.co.uk/wiki/Optimised_contractor_involvement (accessed 17/07/2024).

JCT (Joint Contracts Tribunal) (2016) *Constructing Excellence Contract: Guide*. https://www.jctltd.co.uk/product/jct-constructing-excellence-contract-guide (accessed 11/07/202r).

Kanaris S (2022) Balfour Beatty awarded SCAPE civil engineering frameworks worth £4bn, *New Civil Engineer*, 27 Sept. 2022. https://www.newcivilengineer.com/latest/balfour-beatty-awarded-scape-civil-engineering-frameworks-worth-4bn-27-09-2022/ (accessed 12/07/2024).

Mosey D (2009) *Early Contractor Involvement in Building Procurement: Contracts, Partnering and Project Management*. Wiley Blackwell, Hoboken, NJ, USA.

Mosey D (2019) *Collaborative Construction Procurement and Improved Value*. Wiley Blackwell, Hoboken, NJ, USA.

NEC (New Engineering Contract) (2020) *NEC4 ECC – Practice Note 6: How to use the Information Protocol to support BS EN ISO 19650-2 the delivery phase of assets with NEC4*. NEC, London, UK.

Parker D (2022) Project Profile: How Tideway tunnels were delivered successfully. *New Civil Engineer*, 17 May 2022. https://www.newcivilengineer.com/nce-at-50/project-profile-how-tideway-tunnels-were-delivered-successfully-17-05-2022/ (accessed 12/07/2024).

SCAPE (2020) *Better Procurement. A Response to the Government's Consultation on the Development of an Industrial Strategy.* https://scape.co.uk/research/better-procurement (accessed 12/07/2024).

Smith M (2019) Developments in two-stage contracting and early contractor involvement: A UK perspective. *InfraRead/Ashurst* 13: 36–39.

UK BIM Framework (2019) https://www.ukbimframework.org/ (accessed 12/07/2024).

UK BIM Framework (2020) Information Protocol. https://www.ukbimframework.org/wp-content/uploads/2021/02/InformationProtocolGuidance_Ed1.pdf (accessed 12/07/2024).

emerald
PUBLISHING

ice

Michael Smith, Matthew Finn and Jon Broome
ISBN 978-1-83549-897-2
https://doi.org/10.1108/978-1-83549-894-120241006

Chapter 6
The Client's perspective

Dr Jon Broome

Abstract

This chapter describes

- balancing risk allocation with value for money: what's in and out of the target Prices/Budget
- setting the share profile
- the importance of market consultation
- prescribing a mechanism to arrive at the Prices
- for repeat order Clients: getting a handle on risk; standard documentation
- open-book accounting: what must you see, what do you want to be able to see and focusing on process.
- the difference between monthly verification and periodic audit.

6.1.   Introduction

This chapter is not, strictly speaking, written from a Client's perspective, but as a consultant who regularly works with Clients to develop the commercial and contractual arrangements on which their projects will be let. So, it is really a list, with plenty of explanation, of topics I guide my Clients through in order to arrive at a 'fit for purpose' commercial and hence contractual arrangement. Given that I am usually brought in for bigger, more complex, unique and hence riskier procurements, these tend to involve an early Contractor involvement stage.

6.2.   Be truthful and clear about what your real objectives are

Having understood the rationale for the project or programme in the first place, I often come across bland, imprecise objectives which, when tested, don't stand up to scrutiny.

For instance, repeat order Clients often have a whole portfolio of projects with an annual budget for each programme and state that price certainty is their number one driver for individual contracts. Great, I say, we just need to define exactly what we want, transfer as much risk to the Contractor as possible by using a lump sum Price-based option and deleting as many compensation events (in NEC terminology) that we can, fulfil all Client commitments during the contract (so that there are no compensation events), not change anything during the contract and basically get out of the Contractor's way once it is let. They then explain all the reasons why this will not produce the result they are looking for, including not paying large amounts for risk transfer. This is when the real objectives start to come out.

In the following discussions, it emerges that what they are looking for is value for money (which comes from optimising risk allocation and actively managing the risk, not just transferring it) and foresight on any over- or under-runs, so that they can adjust the programme of projects to match budgetary constraints.

In contrast, for a large one-off project for a regional council, most of the money typically comes from central Government and, if the project over-runs, the already cash-strapped council would either have to find the money internally to the detriment of other much needed services or raise the money from the money markets. It does not want to do either of these things, so wants certainty of price outcome. But if too much risk is transferred to the private sector in order to get that price certainty, as well as having an obligation to provide value for money, the project may well be perceived as too expensive early on – and not be allocated funding in the first place – or the prices come in too high when it is put out to tender or at the end of stage one.

On the other hand, if the project under-runs its budget, the council has to pay the money back to central Government, potentially missing out on their 'wants to have', by which I mean additional features or better quality which saves operating costs. An example of the former is a pedestrian bridge over a bypass which originally was thought not to be affordable. An example of the latter is better quality doors and windows in a new school.

So, for big, externally funded projects, councils want the 'biggest bang for the same buck' and this calls for active management of risk and opportunity, not just transferring downside risk.

This balance of certainty against desire for value for money/least cost also plays out and interacts with
- the desire for certainty of time, such as completion by a set time, against minimum time. An example of the former might be the delivery of sporting venues for a major sporting event. An example of the latter might be the completion of a commercial facility whereby the sooner it is up and running, the sooner revenue will be received
- the scope of the project (in terms of what's in and what's out) and quality (whereby, for example, capital expenditure is largely funded by central Government but the operating expenditures come out of a council's existing budgets).

Often, I have to help Clients get clear on what their priorities are before a 'fit for purpose' contracting strategy can be devised. Given that known, agreed and understood objectives are the number one prerequisite for a successful team,[1] it is worthwhile spending some time on this!

6.3.   Balancing value for money with price certainty and affordability

Having got clear on the above, we can then start to design the 'fit for purpose' commercial arrangement. This will depend on the characteristics of the project and the capabilities and capacity of the likely players, including the Client-side consultants, in terms of what areas of expertise and hence what scope and risk they can take, but also their financial capability to take the risks. Here is my mental process for guiding Clients through the process of developing a fit for purpose contract strategy.

1. Essentially this means developing a high-level work breakdown structure (WBS) for the project, both in terms of what has to actually be delivered and the services that are needed to deliver it. For instance, if large areas of land will need to be compulsorily purchased for a new road, then these form part of the end-product, and a party who has expertise in this area would be beneficial.

2. From this high-level WBS, a package breakdown structure (PBS) can be developed, where each Package is either delivered internally or, if externally, will need a contract. PBS is a concept that the Contracts and Procurement Specific Interest Group of the Association for Project Management (APM, 2017) coined, whereby elements of the WBS are moved in and out of different contract packages to arrive at an optimum breakdown. For example, at main contract level, detailed design is allocated within the same contract as construction. In fact, main Contractors are probably far more adept at this, having to break down the elements of work in their main contract into tens, occasionally hundreds, of supply and subcontract packages.

3. Then for each package, we need to develop the optimum commercial arrangements which are reflected in the individual contract strategy for each package. For simple procurements, such as the purchase of off-the-shelf goods, this will be a price-based contract, without the need for any adjustments to the agreed price, but some warranty and/or remedies for poor performance – for example, late delivery or the goods being defective. At the other extreme – and the subject of this book – for higher value, more complex, unique and inherently risky procurements, you are looking at target cost or alliancing arrangements with early contractor involvement to reduce the risk before some sort of meaningful Price, around which a pain/gain share arrangement operates, is agreed.

4. Having established that the nature of the package lends itself to such an arrangement, the next decision is to decide which of the six early contractor involvement (ECI) types or three alliancing arrangements (or combination) outlined in Chapter 2 best fits the circumstances of the package.

5. It is then a question of refining the selected arrangement by deciding the optimum pain/gain share profile and what's in or out of the fulcrum around which this pain/gain share works – in NEC terminology, this is the target Prices (under Option C of the Engineering and Construction Contract) or Budget (under option E of the Engineering and Construction Contract and the Alliance Contract) – and deciding whether to add new or delete existing compensation events. These are the subjects of the next two sections of this chapter.

One final comment before we move on: while I have described the process in a linear fashion, the reality is that there is a lot of iteration in the above and following sections. Part of that is because major projects tend to have a lot of stakeholders who need to be brought on the journey. Sometimes, they cannot be brought on the journey you want to take them on, so what seems to be the right decision from your perspective is not from theirs and, if they are a key stakeholder, decisions need to be revisited. Alternatively, they give an insight which changes what the 'right' decision is. And included in the stakeholders is the contracting fraternity which we briefly consider in Section 6.7 of this chapter.

6.4.   Principles of risk sharing and allocation

If you talk to lawyers, many will agree that, fundamentally, contracts allocate risk. Yet I have never seen, in any legal text, a discussion around principles for allocating and sharing risk! So, before we discuss the setting of a pain/gain share and what risks, expressed as compensation events in the NEC contract, are deleted or added, let's consider this.

My principles for risk allocation and sharing are explained below.

6.4.1   Consider the effect on the party's business should the risk(s) occur

The extreme of this is that there is little point in allocating a downside risk to a party if it will become insolvent if it occurs: all that will happen is that the risk – and all the other risks! – will ultimately revert back to the Client, but after a long argument during which time the risk is not being managed. A less extreme, but more common scenario, is that the Contractor focuses its efforts on pinning the liability on the Client and vice versa, while all this time the risk is not being managed. Further, this then affects relationships elsewhere, so the management of other interdependent risks (by which I mean those that are best managed with a degree of co-operation) is undermined.

Likewise, we have to consider the effect of cumulative risks occurring. For example, a Contractor might be able to bear the risk of unforeseen ground conditions or bad weather conditions in the earth-moving season or inflation, but not a combination of all three. And this is where the concept of 'sharing' comes in: you allocate enough of the risk(s) to the Contractor to motivate it to manage the risks, but not too much that it will price in an excessive premium or not bid at all. And what is excessive from your point of view might just be a sensible amount from the Contractor's somewhat different perspective (see below).

Now let's consider a typical Contractor's financial strength ... and it's not good.

- On the one hand, as I understand it, the full value of all parent company guarantees has to appear on its balance sheet statement, which superficially makes a Contractor's financial position look very precarious. And in one way it is, because once the parent company is called in due to the Contractor not being able to finance a contract, then it would be like a pack of cards collapsing. However, it would only be called in once the tipping point is reached.
- On the other hand, a small change in costs which the Contractor has to bear has a disproportionate effect on the profit margin. For example, if the Contractor's hoped for profit is 2%, a 1% increase in costs falling on them has a 50% effect on their profit. In contrast, if this cost fell on the Client, it would be a 1% increase in its contract cost. And for the Contractor, a large loss on one contract takes an awful lot of turnover to recover.

You could read the above and think they only apply to downside risk or threat only. But they also apply to upside risk or opportunity. And this is where it works to the Client's advantage: reversing the scenario above, a 1% bonus to the Contractor increases their profit by 50% while only costing the Client 1% of their contract budget.

The last point I wish to make on this heading is that you do have to consider the overall size and hence financial strength of the parties. For example, a £1 m risk to a Contractor with a £10 m turnover is significant and could well cause insolvency but may be insignificant to a £10 bn turnover company. Even though the risk is of the same absolute magnitude, it has a much larger relative effect on the smaller Contractor. Likewise, a Client, such as a statutory undertaker, who knows that it will have hundreds of millions in revenue coming in every year, has a very different ability to bear risk compared with a normal, medium sized enterprise doing, say, a 1 in 10 year modernisation and extension to its factory.

6.4.2   Who can best manage the risk?

Often this is stated as the first and only principle, but I think it comes second. Again, a reminder that risk can be both threat (in which case, we want to manage it out) or opportunity (in which case, we want to manage it in). Given that, for threat, we want ideally not to have it occur at all and for opportunity, we do want it to happen, so all things being equal, we allocate a risk to the party which can best proactively influence whether it happens at all, rather than the party who can reactively minimise the consequences if it does happen. Often these conflict. For instance, the Client can typically best manage the risk of unforeseen physical conditions by doing a good site investigation early in the project's development but, if they materialise, the Contractor controls the resources to minimise the impact.

6.4.3 Clarity over the above, especially for more minor, frequently occurring risks which are allocated within the Prices/Budget

You do not want on-going arguments over whether risks of a similar type are, in NEC terminology, a compensation event or not and, if they are, by how much the Prices or Budget and Completion Date should be adjusted. Rather, you want the parties just to accept that it is within the pain/gain share mechanism and that therefore it is in their joint interests to minimise its impact and hence just get on and manage it, rather than argue over liability.

Under ECI type 4 and Alliance arrangements (as outlined in Chapter 2), this principle is effectively extended so that only major risks and significant scope changes adjust the Budget.

Having talked about the principles, let us now look at the two main mechanisms for allocating and sharing risks, namely the pain/gain share arrangement and what risks are in or out of the fulcrum around which it operates.

6.5. Setting up the pain/gain share profile

The text below is largely taken from my book *NEC4: A User's Guide* (Broome, 2021) and consequently uses the terminology of Options C and D – the target cost options – of the NEC4 Engineering and Construction Contract (NEC, 2017), whereby

- a *share range* is expressed as the range of percentages relative to the adjusted target Price. For example, a *share range* from 100% to 110% would apply to the Defined Cost plus Fees in the range from the target Prices to an over-run of 10%. By 'adjusted', I mean as adjusted mainly by compensation events during the contract. So, it should be noted that the final sharing out of pain/gain is made on the final Defined Cost plus Fee, relative to the final Prices after all adjustments.
- A *share percentage* is the share of any pain or gain that the Contractor takes within the *share range*. So, if the contract has a *share range* of 100% to 110% with a 40% *share percentage*, the Contractor would take 40% of the first 10% of any over-run. If the next *share range* for anything above 110% of the final Prices was 100%, then the Contractor takes all the pain from this point onwards.

Having defined these, exactly the same principles and thought processes apply if there is a Budget (as per Option E of the NEC Engineering and Construction Contract and the NEC4 Alliance Contract) or another form of contract is to be used.

Let us divide the share ranges into five zones, as illustrated in Figure 6.1 and then discussed below.

Figure 6.1 Target cost contract: the five zones (author's own)

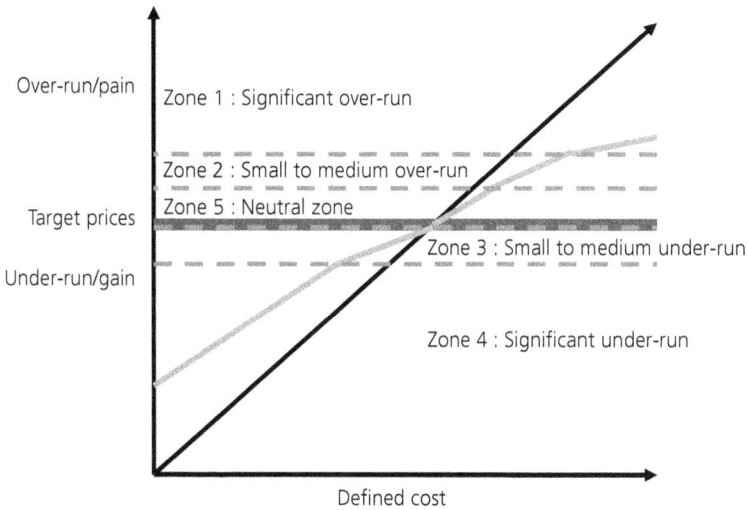

6.5.1 Zone 1: what happens if there is a significant over-run of the target which neither party could have reasonably predicted?

In this zone, the key question is who can best bear the over-run and hence take the majority, but not necessarily all, of the over-run? For instance, a large Client who has money coming in year on year will have much deeper pockets than a relatively small Contractor making relatively small profit on what, for them, is a large contract. If too much risk is put on the *Contractor*, then it will instead fight and squeal to absolve itself of liability to the potential detriment of the *Client's* costs and other objectives. Ultimately, it may become insolvent, in which case all the risk will revert back to the *Client*. If, on the other hand, it is a relatively large *Contractor* (for whom the contract is a small one) and a relatively small *Client* (for whom the contract is a large one), the situation is reversed and it may be advisable to cap the over-run – that is, set it at 100% to the *Contractor*. In other words, the first principle of risk allocation and sharing, as discussed above, has primacy!

6.5.2 Zone 2: what happens if there a small to medium over-run which is within the contemplation of the parties?

Given that risk within the target Prices is generally within the *Contractor's* predominant (but not necessarily exclusive) influence, it suggests that generally the *Contractor* should take the lion's share of any small to medium over-run – for example, a *share percentage* of 50% or more.

The less this is the case, the more the initial pain share should err towards a cost reimbursable contract, with the *Client* taking a greater percentage of over-run – that is, the *Contractor's* share percentage is smaller.

As an example for zones 1 and 2, on the Channel Tunnel Rail Link, the *Employer* had deleted the physical conditions compensation event on large civil contracts, but instead took 75% of any over-run up to 120% of the target Prices (so the *Contractor's share percentage* was set at 25% in my Zone 2) and 90% thereafter (so the *Contractor's share percentage* was set at 10% in my Zone 1).

Likewise, one of my regular training Clients is a water company – a relatively large company with relatively secure income – which mainly contracts with small to medium sized Contractors. Rightly or wrongly, it typically deletes both the physical conditions and weather compensation events clauses to avoid arguments over whether an event is or is not a compensation event and their effects. However, it normally takes the majority of any initial pain share and typically all of it once the Price for Work Done to Date exceeds the Prices by more than 10 or 15%. However, I personally would want the Contractor to still have some 'skin in the game' for larger over-runs, as otherwise it effectively reverts to a cost plus contract with a positive incentive for the Contractor to keep spending!

6.5.3   Zone 3: what happens if there a small to medium under-run which is within the contemplation of the parties?

The mirror image of zone 2 is that, generally, the *Contractor* takes the lion's share of any small to medium savings compared with the target Prices and which are within the contemplation of the parties. Again, the less this is the case, the more initial gain share should err towards a cost reimbursable contract, with the *Client* taking a greater percentage.

6.5.4   Zone 4: what happens if there are significant savings which are beyond the reasonable contemplation of both parties?

This is a point which applies to the first four zones: the share of any savings or over-run for each party should ideally be big enough to motivate both parties to carry on working together to minimise costs.

In one framework let around 2011, the *Employer* – using NEC3 terminology – was consulting with the prequalified Contractors on the draft contract terms, including the share profile, and it was to be an ECI process whereby the final target was negotiated. The *Employer* proposed that Contractors took all the share of any over-run, half of the first 5% of any saving and, beyond that, all savings would go to the *Employer*. As it was in times of a recession, the Contractors were likely to tender low fee percentages/have low margins to ensure they got on the framework. Thinking through the consequences of this from the tendering Contractors' perspective

- they would seek to add in (from an Employer's perspective) excessive risk prior to agreeing the target Prices to protect the downside – that is, their 100% share of any over-run
- they would have no motivation to seek legitimate savings beyond the 5%
- when business picked up in a couple of years' time, the *Employer's* projects would become unattractive, causing them either not to bid or to put the 'C' team on it, having got a nice risk contingency included in the target.

However, in this zone, I see no point in the *Client* paying any more than what is sufficient to motivate the *Contractor* to carry on striving for savings. Equally, giving no gain share for significant savings does not motivate the *Contractor* to find more savings once into this zone.

6.5.5   Zone 5: a neutral zone
In this zone, the Contractor is neither making more profit nor losing its profit. Effectively, the Contractor's share of pain (in Figure 6.1) or gain is set at the same or similar level as the *fee percentage* applied to Defined Costs (using NEC terminology). Therefore, the Contractor is not losing or gaining any head office margin in this zone, although you could argue it is motivated to come in towards the bottom as its profit on turnover figure will be better.

So, why and when would you use a neutral zone? Well, hopefully you will not need to because the parties can agree the Prices. However, if the Parties cannot agree – but are probably within around 5% of each other – then the top of the zone is set at a level that the Contractor can sign up to and the bottom at a level the Client can sign up to. It therefore becomes a tool to avoid protracted and potentially acrimonious negotiation when stage two needs to proceed promptly.

More information and guidance on the working of target cost contracts is given in my other book on collaborative procurement routes (Broome, 2002) or my paper in the *International Journal of Project Management* (Broome and Perry, 2002).

6.6.   What's in and what's out of the target Prices/Budget
As illustrated by the water company example within zone 2 above, the setting of the pain/gain share profile is interdependent with what is in or out of the target Prices or Budget.

Indeed, as per the water company example, the most common changes to that I see to the standard risk allocation of the NEC Engineering and Construction Contract is the deletion of the physical conditions and, to a lesser extent, the weather compensation event clauses, particularly under the target cost options. And the justification for doing this is that they are not directly in the control of the Client (although it can reduce the physical conditions risk by doing a good site investigation), but the Client can take some share of the financial risk through intelligently setting the share profile. However, it should be noted from the

Contractor's perspective that if it is not a compensation event, then the contract Completion Date is also not adjusted, so the likelihood of paying damages for delay increases ... and these are not reimbursable, so fall entirely on the Contractor.

So, a further three comments on the deletion of the physical conditions and weather risks.

- It may be sensible to keep these two risks as compensation events, but with the Contractor only getting additional time and no change to the Prices (and set the pain share accordingly).
- If you are going to delete the physical conditions risk for stage two of ECI, make it clear that the Contractor is entitled and encouraged to do further site investigations in stage one to reduce uncertainty (and set the pain share accordingly).
- Be prepared for discussions around the amount of risk to be included in the target Prices, which may involve adjusting the pain share accordingly. By 'be prepared', I mean put some structure, in terms of contractual words, around how risk amounts are to be calculated. This is discussed in more detail in Section 6.9 of this chapter.

In terms of additional compensation events, the most common I see is an amendment to take account of the re-emergence of COVID or, more broadly, another pandemic. However, prior to the pandemic, the most common additional compensation events normally related to the number of 'business as usual' movements the Contractor should allow for at live facilities. For example, at one port which was being extended from two to three berths, the Contractor was to allow for a certain number of ships arriving and leaving in any month.

Finally, at the time of writing in 2023, inflation is a 'hot topic'. Under the NEC family of contracts, this is not a compensation event. It is covered by secondary Option X1 in the Engineering and Construction Contract but, surprisingly, there is no provision for it in the Alliance Contract.

6.7.   The importance of market consultation

There is no point in developing what, from your perspective, is the greatest contract strategy in the world for your particular project or programme of work, only to present it to the contracting fraternity and for them to refuse to bid. This might be due to entirely presentational aspects, but it is highly likely that they will have some valid points!

In my professional experience, the poster boy (or girl) project for this is the Cross Tay Link Road for Perth & Kinross Council which was given as the example for type 6: optimal ECI with tendered target price in Chapter 2. The 'type 6' gives us a clue in that it is arguably the most developed and innovative type of ECI and consequently would have given the Contractors the biggest surprise if there had been no consultation prior to the tender competition.

Instead, just under a year before the initial prequalification document (ESPD) was due to be issued, the Client held a market engagement day in early March 2019. Following this, four Project Bulletins were issued in the remainder of 2019. The Contract Notice formerly asking for expressions of interest was issued in January 2020 with a draft copy of the conditions of contract as developed and asking for feedback. The ESPD document was subsequently issued at the end of February 2020, together with a significant amount of more detailed information.

At all stages, feedback was invited and reviewed with an open mind, especially if similar feedback was given by more than one potential tenderer. This undoubtedly improved the contract, both in clearing up potential ambiguities in individual clauses and how the contract would work. However, sometimes it resulted in a reiteration or amplification of the same message. Further, feedback on the feedback was given back to the Contractors so they knew that it was not just a marketing exercise by the council.

The result was that a 'full house' of seven Contractors attempted to prequalify. This was reduced to five following prequalification, of which one declined to submit a final bid. Of the four left, it included the one which had perhaps been the most vocal about it wanting a traditional type 3 ECI arrangement with less risk transfer. Further, once let, there have been very few discussions over differing interpretations of the bespoke clauses, as these have been largely ironed out over the consultation and tendering period.

6.8. Prescribing a mechanism to arrive at the Prices or Budget

The biggest drawback of ECI arrangements is the lack of competition to arrive at the stage two target Prices (under Option C or D of the NEC Engineering and Construction Contract) or Budget (under Option E of the same or the NEC4 Alliance Contract). It therefore makes absolute sense to tie down contractually how the stage two prices are to be arrived at.

Unfortunately, even the January 2023 amendments to X22: Early Contractor Involvement of the NEC4 Engineering and Construction Contract are, in my view, deficient in this respect and even if they were not, I would still expect further details to be provided in the Scope documents to make it Client- and project-specific.

So firstly, let's recap from Chapter 2 why I consider the current clauses to be deficient. Clause X22.1(4) of ECC4 states

> Pricing Information is information which specifies how the *Contractor* prepares its assessment of the Prices for Stage Two, and is in the document which the Contract Data states it is in (NEC, 2017a: p.62).

It is specified from Contract Data part two, the part the *Contractor* fills in. A number of points around this.

* Firstly, 'information' implies it is price or cost data filled in and used by the *Contractor* to build up the stage two Prices. And clause X22.3 (5) reinforces this impression by stating that 'the total of the Prices for stage two is assessed by the *Contractor* using the Pricing Information stated in Contract Data'.

However, the Contract Data part two contains other entries related to cost data. What this seems to imply is that Pricing Information is used to build up the stage two Prices, but the other cost data is used for compensation events. Why have two different sets of cost data[2]?

* However, reading the definition a bit more carefully, X22.1 (4) says Pricing Information 'specifies how the *Contractor* prepares its assessment' – that is, a process, not the base data – which conflicts with how Pricing Information is used under clause X22.3 (5). As a Client, I would be unhappy leaving this to the Contractor without some direction from the Client.

* X22.3 (1) states that the *Contractor* prepares its proposals for stage two in consultation with the *Project Manager* and submits them 'in accordance with the submission procedure stated in the Scope'. Does this include for Prices for stage two? If 'Yes' then we have two procedures to comply with: that stated in the Scope and that stated in the Pricing Information, with no precedence given.

Regrettably, this results in me giving the clauses as written a substantial overhaul. Essentially, at a high level, I can break it down into two components: the raw data on which the price is based and the process by which that raw data is used to build up the Prices for stage two.

Let us take the raw data first. Almost certainly, the largest single component of how much the Contractor is reimbursed, under an NEC cost-based contract, is the *fee percentage*. During the contract, this is applied to Defined Costs, whether in stage one or stage two. This is effectively the Contractor's margin – that is, head office overheads plus profit if there is no pain or gain.

At an absolute minimum, to evaluate tendering Contractors' financial submissions we need each one's *fee percentage*. And it makes absolute sense to apply the same *fee percentage* to the Contractor's forecast of Defined Costs at the end of stage two to match how it will be reimbursed during the contract.

I point out that you actually want this to be realistic, rather than as little as possible or too high, for two reasons.

* If it is too small, then the Contractor will be pushing to increase the Defined Cost element of their forecast in order to have a realistic stage two target Price, which includes an allowance for normal profit. At a minimum, this will cause arguments which

can have ramifications once stage two is entered into. In extremis, it could mean the stage two Price is not agreed and a new Contractor needs to be appointed.

- If too large, then the opposite will be true in that the Client may, from the Contractor's perspective, be unfairly pushing down the Defined Cost element of the Price to arrive at a realistic figure. Again, not good for relationships. And if the contract is entered into with an inflated target Prices, the Contractor may not try too hard to reduce Defined Costs as it knows it will be making a good profit without having to find any savings.

Realism is therefore more important than a low *fee percentage*.

For Connect Plus's renewals framework for maintaining the M25 London orbital motorway, as described at the end of Chapter 2, we addressed this point by telling the tendering Contractors what they should allow for profit in their *fee percentage*, which was 2% (a bit below the market rate), because we wanted them to be making profit by generating savings, and gave them a template to fill in for their head office costs. We actually said, 'this is our starter template, but feel free to modify if we have left something out or it does not match how you calculate your head office overheads'. In our tender scoring model if, having audited and verified the build-up, we agreed with it exactly, then the tendering Contractor was awarded full marks – a 10 out of 10 – for that part of the tender assessment. If we were not comfortable with it and it needed modifying, but in a minor way within a preset range, they scored a 5 out 10. If it needed to be modified by more than this preset range, then not only did the tendering Contractor score no marks, but they were eliminated from the competition altogether. This was because we did not want to be working with an organisation we could not trust.

The next thing you want to consider is what, if any, cost data you want the Contractors to tender. The problem is that if they tender a rate for something – a person, a category of person, a type of construction equipment and so on – then under NEC that becomes the rate that will be used, regardless of whether it actually reflects the Contractor's true costs. Consequently, Contractors try to sneak into the contract as many elevated rates as possible as this then becomes a source of 'hidden' profit, both to elevate the stage two target Price and for reimbursement purposes.

There are two strategies for the Client to prescribe here at tender.

- Only allow Contractors to tender input rates which are going to be evaluated as part of their tender submission and ensure that no other rates are inadvertently incorporated into the contract. The rates that are typically tendered are for the staff who will be used in stage one. And to ensure a level playing field when tendering, it is normal for the Client to define what these staff are in terms of job title and description. This might not match a Contractor's internal job titles or how they would like to run the contract.
- Do not ask for them to tender any rates; instead, blitz establishing true costs, mainly of staff, early on in stage one (and ensure that no rates are inadvertently incorporated into the contract).

I prefer the second approach, but many Clients are unhappy with only having a *fee percentage* as the basis for the financial part of the selection.

Now let's turn to process. The inputs, in terms of data, are

- the *fee percentage* and any rates tendered by the Contractor and incorporated into the contract. Under NEC, this means referenced from or stated in Contract Data part two
- rates for the cost of resources which have been verified in stage one – see the next section of this chapter – and which may well already been used and hence paid for as Defined Cost. This could be staff costs or hire costs of construction equipment used for enabling or advanced works in stage one
- subcontractor quotations for the subcontract packages.

The neatest way to do this contractually would be for the Client to specify a high-level process in its Scope and then ask the Contractor, at tender and as part of the quality assessment, to develop specifics of how it would comply with the Client's requirements, which are then incorporated into its Scope and become a contractual obligation – that is, not to have a separate Pricing Information document. And this is what I used to do, with all references to the Pricing Information deleted.

Having said this, with Clients who know from experience how they want the target Price or Budget built up and the capabilities of Contractors to do it, they have prescribed in a fair amount of detail both the process and the end result, in terms of what is in the final submission.

And that would bring us very nicely on to open-book accounting and the importance of getting it up and running early so that the data used to build up the stage two Prices/Budget is the same as that which will be used to reimburse the Contractor. However, perhaps the most difficult aspect to agree on is the quantification of risk for inclusion in the Prices/Budget.

6.9.   Getting a handle on risk

From experience, this is the most contentious area when agreeing the target Prices. Why? Because, in my opinion, the 'science' of quantitative risk analysis has been based on the subjective assessments of risk, particularly probabilities, where the size of the data sets – individuals' experiences – are too small to be statistically valid and are riddled with the cognitive biases which we have as human beings – in other words, 'garbage in' to a mathematically sound process still results in 'garbage out'. Consequently, Clients are either hood-winked by the Contractor (who manipulates the input data to give a beneficially high contingency for inclusion in the target Prices/Budget) or both parties are engaged in guesses to come up with the figure.

For the Connect Plus hybrid example given at the end of Chapter 2, the amount of risk contingency to include in the Budget for the package was the hardest thing to agree, especially as it needed to include risk allowances for what would normally be compensation events. This was despite

- work of a similar nature in similar circumstances had been done under the previous framework so Connect Plus, as the Client, 'should have' had the data available
- as it turned out, the same Contractors being selected for this framework as for the previous framework, so they also 'should have' had the data readily available.

In reality, it was hidden in numerous individuals' Excel spreadsheets for the contracts going back the previous four years. And even if we found it, it then had to be applied in the context of the current physical works in any package and the commercial relationship. The truth is it took about two years of collecting data from the current framework before we could start to get an objective, as opposed to subjective, handle on the 'right' amount of risk contingency to include for each package.

So, if it's hard get a handle on repeat order work of a similar nature, then it's going to be even harder to get a handle on risk for novel, complex, one-off projects of the type where an ECI arrangement is typically going to be used.

Consequently, as part of the Scope for how the target Prices/Budget are arrived at, I would want to describe the approach in the Scope, perhaps with the Contractor fleshing it out in their quality submission. This is then incorporated into the contract as, using NEC4 terminology, the *Contractor's* Scope. And included in this would be what databases and tools were to be used.[3]

6.10.   Nailing open-book accounting from day one

Getting open-book accounting up and running from early on in the contract is one of my five critical success factors for a successful NEC Engineering and Construction Contract and the only one which specifically applies to the cost-based options C to F.[4] It becomes even more relevant and important under an ECI arrangement. Why is this? Let's take the case of a plain cost-based contract without ECI first.

Over the years I have seen a huge disparity in how target and cost reimbursable contracts have been administered, to take a couple of examples at one extreme.

- The contract became a bureaucratic quagmire of pettiness with, in once case, the Employer's (as it was under NEC3) quantity surveyor (QS) insisting on being supplied with a physical copy of every subcontractor and supplier tender inquiry, order, invoice, delivery note and proof 'beyond reasonable doubt' that it had (past tense) been paid and disallowing cost if he thought the Contractor was paying more than the lowest cost he could source. This goes way beyond the requirements of the NEC contract and wider contract law, both in terms of the evidence required and demonstrable value: Defined Costs have to be at 'open market or competitively tendered prices' (NEC, 2017a: clause 52.1), not lowest cost. Further, the Employer's QS was somewhat zealous in the application of other Disallowed Costs. And the Employer was paying thrice for this bureaucracy: once for the Contractor to bring forward and supply all this evidence, once for their own QS to evaluate and again for the arguments. Ultimately, the Contractor went to adjudication because it was seriously under-recovering and the Employer decisively lost the adjudication.

- On one big project of several £100 million to significantly upgrade and expand an underground station, the Project Manager's commercial team and Contractor had been happily agreeing Defined Costs as the contract progressed. Unfortunately, in the later stages when the mechanical and electrical systems were being installed, costs started to run away, with the Defined Costs plus Fee – the Price for Work Done to Date in NEC terminology – exceeding the target Prices. It was only then that the Employer decided to do a 'proper' audit of Defined Costs by calling in a separate firm of forensic auditors who were paid on the basis of a share of the saving they found: hardly the 'impartial administrator' of the contract as required by the first case law on NEC.[5] Unsurprisingly, given their natural bias, they interpreted the contract at the extremes of interpretation in the Client's – and their – favour (which is not to say that some of the issues they found weren't valid, but wouldn't it have been sensible to pick them up early in the contract?). The net result was that the Client wanted midteen millions of pounds back from the Contractor, who was naturally unhappy about this. Again, it resulted in a big adjudication, which the Client substantially lost.

At the other extreme, well, a relatively hassle-free process!

The above is a big enough reason to nail this on any cost-based contract. On an ECI contract, you also want the cost data that is used to build up the stage two target Price/Budget to match how the Contractor will be paid. Otherwise, as stage two progresses, even if the contract is exactly on-track, disparities between the forecast and actuals will open up.

The good news is that, using an ECI approach, you can nail this early on, both before the target Price/Budget for stage two starts to be built up in earnest and before the big money starts to be spent in stage two. So here are my high-level top tips, again using NEC4 terminology[6].

6.10.1 Pre-verify, audit and/or agree as much as you can before any Defined Cost is incurred

By this I mean, for example

a) the People rates which will be charged per unit of time[7]

b) the level of detail that is required to justify that a person has spent time working on your contract, rather than another one or somewhere else in the Contractor's business.

My experience is that the consultants, who are typically Subcontractors to the main Contractor, like to charge all their time to the contract with little justification of what they were actually doing

c) the rate per unit of time for Contractor-owned construction equipment

d) the processes by which Subcontractors and suppliers of construction equipment and plant and materials are selected so that the Client can be confident that the rates are 'open market or competitively tendered prices' (NEC, 2017a: clause 52.1)

e) what the head office rebates and discounts are, which may or may not be applied to the previous point (d), and whether they are applied before or after they appear at the contract-level interface

f) the various processes by which a cost or time arrives at the Contractor's team's terminals. For example, if the Contractor hires in some construction equipment, when does the hire start and how does the data appear at the Contractor's terminal? Likewise, when it is off-hired – does that off-hire start at the point off-hiring or when it is taken off site? With or without any discount, as per point (e) above? The same could be said for site attendances for People and Plant and Materials. Understand the process and you understand the cost that pops out the end. Further, when you do subsequent audits as per Section 6.10.3 below, errors are viewed as areas for improvement in the system rather than being taken as attempts to over-charge

g) how Disallowed Cost will be managed (another process). What you do not want is a bow wave of disputed Disallowed Costs building up and getting progressively bigger as the contract progresses. The Contractor will typically then submit the same disputed amounts every month for payment and the Project Manager's commercial team will disallow them every month. This is extremely wasteful of everybody's time. I suggest that any disputed Disallowed Costs are separately identified, with actions identified to resolve the dispute: many will just be that more information is needed. Once the action is done, it can be removed from this list and then never needs to be revisited again as a possible Disallowed Cost.

Just to re-emphasise the point I made at the start: pre-verify, audit and/or agree as much as you can before any Defined Cost is incurred!

6.10.2   Establish what is needed on a monthly basis

a) by the Client, forwarded by the contract commercial team every month, for its permanent records. Particularly for public sector Clients, audits might be done by separate bodies several years after the contract has ended. If there are no records, then there will be no means of justifying the amounts paid apart from reference to market rates which is not Defined Cost

b) accounts, records and data that the Project Manager's commercial team[8] requires to see every month, which relates to what checks they want to do every month for each application and certification

c) accounts, records and data that the Client's commercial team want to be readily available to see. This could be because they want to do deeper dives to verify the data which appears on the higher-level accounts and records. These could be done on a periodic/rotating basis or a risk-based approach, which could be because something doesn't match-up (including it has cost more than forecast) or it is a high-value item.

d) the overall monthly process by which the Contractor builds up and submits its application and the Project Manager certifies it. There are two things to note.

- o I view it as a monthly process, not a point in time, so while the 'what' has been identified in points (a) to (c) above, 'when' is the Project Manager's commercial team going to have access to the different accounts, records and data identified?
- o What are the interactions/meetings over that month when details and clarifications are worked through?

The end result should ideally be that the Contractor's application is the same – as what the Project Manager certifies as everything has been progressively agreed.

There should be a degree of 'meeting of minds' on the above points because what you do not want to do is impose overly onerous criteria on the Contractor whereby it spends significant time – at the Client's expense – manipulating its internal accounts, records and data into the form favoured by the Project Manager's commercial team or retrieving minutiae from the depths of its accounting system which is rarely or never looked at.[9] Ideally, no manipulation is necessary.

6.10.3 External audits

a) How often are interim audits by an external or more specialist team going to come in and on what basis are the auditing different things: random, risk-based and/or rotating?

b) Will it be determined in consultation with the permanent commercial team or be completely independent? What is the composition of this team? For instance, having given training to Client's accountants on the detail of the Schedule of the Cost Components, their comment was that you really needed a specialist payroll accountant to properly verify People rates in accordance with the Schedule.

c) Will it be the same team involved in the initial setting up? If there has been a strong emphasis on establishing good process, then using the same team can confirm that the processes are working in practice while identifying any areas for improvements.

d) How will details of the audit be arranged? It can be argued that too much prewarning allows the Contractor to 'stage manage' it but, at the other extreme, you do not want a team of auditors turning up unannounced at a Contractor's head office demanding access to confidential records!

6.10.4 Final accounting

What else do you expect to see for 'final accounting' as part of Defined Cost in NEC4 ECC clause 50.9 of options C to F? If the previous have been done properly, it should be minimal and final accounting should be a formality. And that is all I am going to say on this point because it should be that simple. If you don't do the above, then potentially there will be a lot more to say: see my two examples of bad practice at the start of this section!

Intricately linked with the process of 'how much' the Contractor is paid every month is accurate financial forecasting to completion of the contract. That links in with progress achieved and value delivered and so on, so I will not say more about it apart from do not forget about it. The foundation for knowing where you stand currently is understanding how the Price or Budget was built up, including for risk (see Sections 6.8 and 6.9), and setting up the open-book accounting correctly as per this Section.

Whatever is agreed should be recorded as Contractor's proposals for stage two which then, following the notice to proceed, becomes a contractual obligation on all parties.

The result of 'front-loading' the process is that there should be far less hassle for both parties – which means far less cost for the Client as it is paying for all of this – and both parties knowing where they stand financially as the contract progresses. Consequently, all participants can focus on managing the works instead of arguing over payment.

6.11. Conclusions

To sum up this chapter, I would say 'the devil is in the detail' or, more precisely, the specifics. By this I mean

- be specific, as well as truthful and honest, about what your objectives for the project and hence contract are
- be specific about how risk is going to be allocated and shared through additional or deleted compensation events and the specifics of the pain/gain share mechanism. These decisions need to be thought through from both the Client's and Contractor's perspectives, taking account of not just ability to manage but effect on the parties' objectives
- communicate and get feedback from the market: even if you do not take it on, communicate why so they understand where you are coming from
- be specific about how the target Prices/Budget for stage two will be built up and presented. This includes for the most contentious area – that is, risk amounts
- be specific with the Contractor early on in stage two about how Defined Costs will be built up and used both for setting the target Prices/Budget for stage two and for reimbursement. For the latter, ECI gives you the time to nail this: view the time taken to set this up as an investment which will save both parties hassle, time and cost (for which the Client is ultimately paying) and avoid a potential dispute.

Notes

1. Although published some 20 years ago, Hackman, a professor at Harvard University identified this from all the research that he is aware of as the number one factor! See Hackman JR (2002).

2. Related to this, in a consultancy assignment for an up and running framework which involved using ECI, there were effectively three sets of pricing information and rules for their application: one to get onto the framework, one for pricing each contract and another for reimbursing the Contractor during a contract, none of which was clearly delineated in the structure of the contract documents. You can imagine the confusion this brought to the administration of the framework and individual contracts. You don't want this! Instead, you want the cost information and how it is used to build up the Prices/Budget to match how the Contractor is reimbursed during the contract.

3. It may well be that the Contractor, who has done similar projects, has data which can be used. Alternatively, databases and tools, such as nPlan's which has a huge database of past projects to draw on and uses artificial intelligence to select the appropriate risk and variances, can be used to take out much of the guess work. It can also help de-risk the contract by highlighting the riskiest areas to focus on. Author's note: I do have a commercial relationship with nPlan.

4. I could describe the other four but then this would become a book on the NEC4 Engineering and Construction Contract.

5. *Costain Ltd* v. *Bechtel Ltd* (2005), which reflects how contracts are interpreted generally i.e. as an 'objective third party' would. The decisions of a consistently biased Project Manager would not stand up to this test if a dispute gets to such a person e.g. an adjudicator, arbitrator or judge.

6. I could go into detail on individual headings in the Schedule of Cost Components, but then this would become a book on NEC rather than Early Contractor Involvement.

7. Under NEC, if the Contractor has included in or referenced People rates from the Contract Data, then you have accepted these as part of its offer and they cannot be reviewed. However, additional rates for new People can be audited for build-up in accordance with the contract i.e. People of the Schedule of Cost Components.

8. I say the 'Project Manager's commercial team' to emphasise the point I make in footnote 5, namely that all these actions fall on the Project Manager under NEC and that he or she is an 'impartial administrator'.

9. This is a good example of where understanding the process pays dividends: understand once how this low-level data is assembled and then collated into the higher-level information that is presented. You can then largely forget about it for the monthly application and certifications, which is not say the process is not occasionally audited to confirm it is working as expected.

REFERENCES

APM (Association for Project Management) (2017) *APM Guide to Contracts and Procurement for Project, Programme and Portfolio Managers.* APM, Prices Risborough, Bucks, UK.

Broome JC (2002) *Procurement Routes for Partnering: A Practical Guide.* Thomas Telford, London, UK.

Broome JC (2021) *NEC4: A User's Guide.* ICE Publishing, London, UK.

Broome JC and Perry JG (2002) How practitioners set share fractions in target cost contracts. *International Journal of Project Management* **20(1)**: 59–66.

Costain Ltd v. *Bechtel Ltd* (2005) EWHC 1018.

Hackman JR (2002) *Leading Teams: Setting the Stage for Great Team Performances*. Harvard Business School Press, Boston, USA.

NEC (New Engineering Contract) (2017a) *NEC Contracts*. NEC, London, UK. https://www.neccontract.com/ (accessed 09/07/2024).

NEC (2017b) *NEC4: Engineering and Construction Contract*. NEC, London, UK.

Michael Smith, Matthew Finn and Jon Broome
ISBN 978-1-83549-897-2
https://doi.org/10.1108/978-1-83549-894-120241007

Chapter 7
The contractor's perspective

Matthew Finn

Abstract

This chapter describes

- the risks for a Contractor associated with early contractor involvement (ECI), particularly when budgets are not yet defined for the scope of works
- the benefits of ECI to identify risks and share/price those risks
- the costs and other disadvantages to the Contractor of ECI
- the opportunities gained through ECI to enhance the value of the design, clash coordination and value engineering
- the issues and barriers to early involvement of a Contractor and unrealistic expectations for the Contractor
- the stage at which the Contractor should be introduced – when is it too early for ECI?
- the commercial considerations, risk allocation and incentives.

7.1. Introduction

In this chapter, we focus on early contractor involvement (ECI) from the Contractor's perspective.

7.2. The risks for a contractor associated with ECI

Stage one of ECI allows the Contractor to get a clearer understanding of the Client's expectations. ECI encourages the project team to focus on the Client's objectives from the outset; this in turn can result in a better outcome for the Contractor, allowing substantially more time, depending on the length of stage one, to identify, avoid or manage construction risks (NEC, 2020). ECI encourages the project team to focus on the Client's objectives from the outset (NEC, 2022). This can result in a better outcome for the Contractor, allowing substantially more time, depending on the length of stage one, to identify, avoid or manage construction risks (NEC, 2022).

During stage one the parties may discuss ways to mitigate and minimise the risks that may occur during stage two – for example, the project team can work together

▦ to review and test proposed construction methodologies, programme and cost estimates before construction (NEC, 2020). Reviewing the cost estimate and the programme prior to commencement of construction minimises the risk of conflict between the parties on the cost of activities and unachievable programmes

▦ to decide on how the works should be procured for subcontractor packaged works which can have a positive impact on the Contractor's cash flow. The Contractor can use its knowledge on how to package up the works in a way that results in the programme showing certain activities being carried out in particular periods so that it can design a programme that gives them a desired cashflow; for example, it may front-load the programme, so that particularly expensive items are paid very early on in the programme

▦ to assist with the hiring of specialist subcontractors, giving the Contractor the opportunity to influence the procurement by providing recommendations to the Client on preferred specialised Contractors that the Contractor has a great working relationship with; this will have a positive effect on collaboration.

The level of risk associated with the Contractor's scope of works will vary depending on the timeline of the ECI's phases and the development of the design, maturity to define the scope of works, political influence, delayed decision making and changes to industry and technical standards. These impacts can prolong stage one of the ECI beyond the initial pro-gramme and may require additional funding to be committed to the project, an issue often seen in mega projects such as Hinkley Point C, High Speed Two and so on. The delays to stage one often impact the Contractor's costs associated with undertaking subsequent periods of 'optioneering' and affordability reviews. The risk to the Contractor is that it generates an overspend on anticipated resources allocated to stage one, which may prove difficult to recover, particularly when the Client is offering a large volume of upcoming revenue to do the project. In the event that the project does not later go ahead, this can leave the Contractor with an under-recovery on stage one and a large hole in its future order book, which can give rise to a dispute surrounding the costs of stage one and the Contractor's entitlement to recover its cost for variations relating to value engineering, optioneering, affordability reviews and so on.

7.3. The identification of risks and sharing/pricing those risks

The parties identify risks together as the project develops prior to construction commencement. The parties can discuss ways to mitigate and minimise the risk – for example, the project team can work together to optimise buildability, which gives the Contractor an opportunity to share knowledge and suggest build methodologies which are easier and safer to eliminate safety risks for the Contractor to manage on site. The risk register can also identify risks to prevent design changes during construction, which is when design changes are more expensive, and may result in prolonging the construction programme. Identifying potential design changes earlier makes the changes less disruptive to the Contractor's construction works.

ECI enables the project team to review and test proposed construction methodologies, programme and cost estimates before construction (NEC, 2020). Reviewing the cost estimate and the programme prior to commencement of construction minimises the risk of conflict between the parties on the cost of activities and unachievable programmes.

Having a robust and transparent method of pricing and allocating risks between the Contractor and Client can give the Contractor better risk protection to protect its fee and minimise its exposure. Allocating risks as soon as possible gives the opportunity to mitigate the impact of such risks. This is a key feature of ECI that allows the Contractor to contribute to the design development and planning of the construction phase. ECI also provides a key opportunity for the parties to foster innovation and promotion of a collaborative approach within the project team. This is vital in complex engineering projects which require solutions to key logistical and buildability risks. The Contractor can add solutions such as lean construction and off-site manufacturing to overcome such difficulties. This not only provides a benefit to buildability but can also speed up planning and constraint issues with third parties whereby consents are required preconstruction.

Other key areas in which collaboration with the Contractor can define and pass risks on mega projects through ECI are as follows.

- The Contractor can advise on the buildability of the design, sequencing and programming and the best way to mitigate risks in construction.
- The Contractor can develop site layouts and locations of site accommodation that enable the Client to start negotiations on any land-take agreements in advance of the works starting in order to avoid delay.
- The Contractor can advise from its supply chain lead-in times for key materials and equipment as well as develop decisions as to who is best placed to order these to maximise efficiency in the programme.
- The Contractor can provide an 'open-book tendering' opportunity to negotiate on price and commercial terms with its supply chain, providing the Client with full visibility as to how the price is built up and allowing them to see that a robust procurement exercise has been completed.
- The Contractor can identify potential risks with coordination of the design and statutory utilities diversions to minimise/avoid clashes with utilities and structures and so on.
- The Contractor can input into the methodology of the works to ensure that the risk to health and safety is removed or reduced.
- The Contractor can find, with its supply chain, efficiencies in the programme for the works by finding reductions in Contractor's indirect costs, thus mitigating the risk for future delay.

However, the Contractor undertaking this role in an ECI will be taking on risk from the Client. This will provide the Client with a higher upfront cost and can give rise to many disadvantages, such as the concept of a 'lock-in' of the Contractor into the future of the project.

7.4. Costs and other disadvantages to the contractor of ECI

It is to the Contractor's benefit to ensure that it has identified all the risk and scope of the project, as far as practicable, in the early stages of the ECI. This will minimise the risk of any future dispute surrounding whether the Contractor has missed an element of work or risk in the target cost and thereby lost its entitlement to a change or an effect on the pain/gain mechanism. This depends on the ECI agreement – that is, if there is a specific pain share mechanism; if it is a moderated fee the cost will still be paid to the Contractor. However, in a more traditional arrangement, this could result in a measurable unrecoverable cost.

Therefore, an important commercial consideration for a Contractor in an ECI is to get the price right to set budget and incentivisation budget. Whether the Contractor gets this right will depend on the maturity of the design, the appropriate risk allocation and the Contractor's ability to manage those risks within the Client's target budget.

If too much risk is passed to the Contractor during the ECI this then can exceed the budget for the project and gives the heightened risk that the project may not be viable or, if the project progresses, whether the Contractor can sustain the passed-through risk and appropriately complete the project given the commercial pressures. It may be more sensible for the risks to be a 'neutral' risk or for the parties to continue to find a relevant solution as the project commences. Under NEC4 ECI, if the Contractor does not agree with the Client on how to resolve issues in stage one, such as agreeing to a stage two price and or agreeing to a Completion Date, then the Contractor can decide not to proceed to stage two, at which point the project manager must issue an instruction removing the issue in disagreement from the scope of work (NEC, 2020).

There are further downsides in ECI, such as if the Contractor through a lack of experience is not managing an early warning procedure and notifying the Client of its escalating costs, this can cause the Client to feel as if it is being railroaded into an ever-increasing budget or high price.

With ECI the correct commercial and contractual model needs to be adopted with the right balance between incentivising the Contractor to engage in driving solutions and for the Contractor to see the clear benefit in doing so. This will avoid the pitfall of the Contractor driving the budget up as high as possible through the identification of issues and simply adding these to either the risk register or the budget.

Therefore, the timing of the appointment of the Contractor in an ECI is key in consideration to testing the pricing of the market. In an ECI where the Contractor assists with the design, it means that it is seen to be intrinsically linked to the project when other competing Contractors might not be interested in bidding for the project if a two-stage tender procurement method is adopted. Having a new Contractor take on the design from the ECI Contractor could incur additional risk/cost on the project as there would need to be additional time for the new Contractor to validate the design to date.

7.5. Opportunities gained through ECI to enhance the value of the design, clash coordination and value engineering

Coordination, or specifically design coordination, on a construction project is the process of checking each party's information against the rest to ensure a single, unified set of deliverables that achieves the project brief. Coordination has always been paramount to the success of any construction project. Without robust coordination there is a great risk of errors, increased costs, programme delays or Client dissatisfaction.

In design terms, a clash occurs when components that make up a built asset are not spatially coordinated and conflict. The engagement of the Contractor and key members of its supply chain during the early design stages can improve clash detection and bring potential value by highlighting significant issues to the project team earlier in the project programme, avoiding potential costly changes to the design after packages have been procured or on site. Rectifying errors at the design stage is easier and much less expensive than dealing with those errors during construction on site or in the factory.

7.6. Stage for the involvement of the contractor

The integration of experience and construction knowledge into the early phases of a project is of most benefit. The earlier phases of a project are characterised by the greater potential to influence the design with a lower impact on cost.

In 1964, the argument for appointing the main Contractor early was articulated clearly by Banwell who stated that there are

> occasions when it is appropriate for the main contractor to be appointed and brought into the team before the design is finished and the programme of work finally settled (Banwell, 1964: p.4).

Banwell also recommended the early appointment of certain specialist Contractors as design team members in addition to early main Contractor appointments (Mosey, 2009).

The Client's procurement strategy will determine when the Contractor will become involved in the project. In most standard building contracts, the Contractor is procured relatively late in the process – for example, at RIBA Stage 4: Technical Design, when key decisions have already been taken by the Client and design team on design and construction methodology. Introducing the Contractor at an earlier stage – for example, RIBA Stage 3: Spatial

Coordination – would enable the Contractor and its supply chain to utilise their specialist skills and knowledge to help drive greater efficiencies.

The RIBA Plan of Work 2020 recognises that for some projects, information confirming buildability, site logistics and other construction-related matters, such as the adoption of modern methods of construction in the construction strategy, will be more comprehensive and robust if the Contractor is appointed and involved at RIBA Stage 2: Concept Design (RIBA, 2020).

In 2014, a conference paper which discussed the advantages and disadvantages for the design team of using ECI (Sødal *et al.*, 2014) identified that even though the collaboration between designers and construction personnel can be productive, the inherent conflict of interest can threaten a positive outcome. The Contractor tends to have a strong cost focus, which may result in a simple design and familiar solutions rather than creative and innovative solutions, whereas architects tend to prefer a separate design as they typically believe quality is better maintained in that way.

So, the main barrier is therefore to ensure that every participating party agrees on a united goal and pursues this in a collaborative manner (Sødal *et al.*, 2014). To create an effective and efficient working environment there needs to be an effort to form relations, identify the participants' abilities and nurture mutual respect among the participants; this will help reduce the number of misunderstandings and unnecessary conflicts (Sødal *et al.*, 2014).

ECI has been used by many public sector construction clients; however, some clients have been reluctant to adopt the strategy because of concerns associated with appointing Contractors before permissions such as planning have been granted, and issues to do with attracting accurate and competitive prices from Contractors when the design is not yet well advanced (Designing Buildings, 2020).

NEC (2020) published a practice note on ECI contracts suggesting that the proposed timing of the ECI Contractor appointment in the project lifecycle has an impact on all aspects of ECI contract development and procurement. The ECI contract is most beneficial for clients that appoint the Contractor before planning procedures are completed as this provides the opportunity for better schemes to be developed and submitted for planning approval. However, ECI awards prior to completing the planning procedures may require more assumptions to be made in the production of the requirements and budget, so it is a balance of risk and opportunity. An early procurement planning task should be to decide whether ECI is suitable for the circumstances and, if so, the optimal timing of the Contractor's appointment. It is good practice to consult with the market regarding the key aspects of the proposed approach, including the timing of Contractor appointment, contract options and procurement schedule (NEC, 2020).

The NEC practice note on ECI describes the importance of the budget being prepared by the Client and states in the contract data that

> ...it is essential to the mechanics and ultimate success of ECI using X22 that the Client includes an achievable Budget within the Contract (NEC, 2020: p.2).

If the Client appoints several Contractors for ECI during stage one, this can make collaboration difficult for the Contractor. Appointing several Contractors may reduce the incentive for them to work collaboratively and openly with the Client team, thus reducing the benefits expected for an ECI process (Cheung, 2019).

The budget may be more difficult to determine for projects where the Contractor is to be appointed very early (NEC, 2020). The earlier the Contractor is involved, the higher the level of risk there could be and the less accurate the budget is likely to be (Kan, 2018). Where the budget is not defined, this means that the budget does not include an allowance for both the Contractor's risk and the Client's risk and inflation through to the end of the contract. Where there is no budget, there is potential for conflict between the Contractor and the Client regarding expenditure, costs and estimates for works.

During the procurement stage of appointing the Contractor for stage one, the Contractor will identify commercial risks when tendering for the works. Once the Contractor is appointed to begin stage one, the Contractor may identify more risks than it envisaged during the tender stage; those risks identified during stage one may result in the Contractor being less likely to develop a project within budget (Kan, 2018).

If the Contractor submits a price to complete the stage two scope of works that is higher than the Client's budget, NEC ECI X22.5 (3) (NEC, 2020) allows the Client to elect not to proceed with stage two or to engage another Contractor to undertake the stage two works.

If the Client decides not to continue with stage two with the same Contractor from stage one
- the Client will not be liable to the stage one Contractor for loss of opportunity or profit for the omission of the stage two works where the parties are unable to reach an agreed total of the prices for stage two or if the Contractor fails to achieve the required performance (Cheung, 2019)
- the Contractor might be reluctant to transfer significant intellectual property to the successful Contractor (thus eroding the losing Contractor's commercial advantage on future projects) (Purcell, 2022).

These barriers restrict the Contractor's negotiating position during establishing the prices for stage two because the consequence of not being given the work would impact the Contractor's profit and resource allocation as well as loss of opportunities to work on other projects and loss of intellectual property; therefore the Contractor may accept lower prices than it would like to prevent the consequences of not being awarded stage two. The risk of agreeing to a lower price may also mean that the Contractor is at risk of the final project cost exceeding the budget and therefore has no entitlement to a budget incentive (NEC, 2017: clause X22.7(1)).

If no agreement can be reached on the change, the project manager is responsible for its assessment, as set out in clause X22.6 (NEC, 2020). Before making this assessment, the project manager should first discuss the matter with the Contractor and the Client (NEC, 2020). The project manager must issue an instruction removing the contentious issue from the scope of work (NEC, 2020). Where the project manager removes the scope of work due to budget insufficiencies, this may have a negative impact on the Contractor: reducing profit margins, missing opportunities to tender for other projects and difficulties relocating planned allocated resources.

Alternatively, if it becomes evident that the budget is not sufficient for the project, clause X22.6 allows for changes to the budget to be made. However, changes are only made because of an instruction changing the Client's requirements stated in the scope of works. Accordingly, the Client's requirements need to be carefully drafted to identify those matters which are essential (NEC, 2020). Therefore, if the scope of work is unclear, this may result in conflicts between the Contractor and the Client as to what constitutes a change to the Client's requirements. The Client's requirements should be clear, concise and provide adequate detail of what the Client expects to be achieved with the budget. The information may include general information about the project goals and objectives, relevant investigation and survey reports, early-stage design drawings, identification of risks and restrictions that may impact the works and so on.

ECI requires that the budget is established prior to the appointment of the Contractor for stage one. If the Contractor is appointed prior to the budget being established; the Contractor will work with the Client during stage one to try to establish a budget so the project can be developed sufficiently to begin stage two.

ECI requires a collaborative mindset to make it successful for all parties. If the Contractor is appointed prior to a budget and risk allowances being established, this will make project delivery harder and potentially increase conflict. Collaboration can be easily affected by too many people being involved at stage one. The impacts of cash flow, design efficiencies and risk of design changes are improved through NEC ECI X22.

7.7. Case studies: benefits and opportunities gained through ECI

Case study – Hinkley Point C

In 2014, prior to the commencement of the Hinkley Point C, a £33 bn new-build nuclear power station, the operator EDF Energy engaged in workshops (EDF Energy, 2014) to learn from the construction of the European pressurised reactor (EPR) nuclear power plants (NPP) at Flamanville NPP in France which was €19 bn, and Taishan NPP 1 and 2 in China which was US$7.5 bn.

The construction of nuclear power plants often involves complex engineering, regulatory compliance and safety considerations, making it susceptible to budget over-runs. Several factors can contribute to these cost over-runs, such as regulatory changes, design changes, supply chain issues and so on. New-build EPR reactor designs have often seen large cost over-runs such as Olkiluoto 3 NPP in Finland which was 12 years late and suffered cost over-runs of €3 bn to around €12 bn. Similarly, Flamanville NPP is yet to be completed but was scheduled for completion in 2012 and has cost over-runs from a budget of €3 bn to c. €12 bn. This history of the EPR and the further gloomy study from Massachusetts Institute of Technology (MIT) of over 50 years of US nuclear power plant construction data show that building plants based on existing designs (i.e. EPR) actually costs more, rather than less, than building plants based on new designs (Chandler, 2020).

Therefore, Hinkley Point C had to address key procurement lessons learned from these previously built EPRs as its success and the success of Sizewell C is vital to the success of the UK's renewable energy platform. Hinkley Point C uses a mixture of contracts for the procurement of its Contractors using FIDIC 1999 Yellow Book (Plant and Design Build) (FIDIC, 1999) on around 150 of the 300 + contracts (Walker, 2022); however, the majority of the contracts are NEC (NEC, 2016).

The key lesson learned from the workshop was that ECI improves constructability and interface management. Therefore, EDF developed a procurement strategy of having an Early Work Agreement to cover key areas of the project such as

- critical engineering activities
- technical configuration and interface management
- project management activities ramp-up
- plot plan finalisation/optimisation
- site preparatory work
- support to licencing
- schedule.

Other key findings and benefits were finding the appropriate Contractor methods at the very beginning of the design to improve constructability. An example of such a benefit was the prefabrication of the pools for the reactor, which was subsequently adopted into the project and became a key milestone (Moore, 2023).

Due to the complex nature of a new-build NPP, ECI is vital to the success of the project as the technology needs to be selected and therefore the relevant selection of turbines, generators and so on all needs to feed into the early regulatory approvals and consents to be obtained and the subsequent long lead-in periods for procurement of such equipment. The benefit therefore of having a Contractor appointed through an ECI is that this process can begin to develop a design which addresses buildability issues while coordinating these with key specialist suppliers.

Case study – High Speed 2

High Speed 2 (HS2) is a new-build connecting 140 miles of rail infrastructure between the UK's two largest cities, Birmingham and London, with the option to extend to Manchester and the North West as well as Scotland. The project started in 2017 and has already seen much controversy, with the original budget set in 2009 at £37.5 bn but, as of 2021, the Department for Transport's estimate is between £72 bn and £98 bn.

The project has presented many engineering challenges which required ECI to a joint venture of Balfour Beatty and VINCI (BBV) to advise on the buildability of the design, sequencing and programming and the best way to mitigate risks in construction. This has given the benefit of BBV and HS2 working together on identifying and resolving issues that could affect construction – that is, de-risking a particular constraint or appeasing a third party whose consent is needed.

A good example of this is the Marston Box Rail Bridge (HS2, 2023) which is the world's longest box bridge slide across the M42 motorway. BBV was engaged under stage one of the ECI to undertake early engagement with HS2 and National Highways to obtain the consent of the box slide solution and minimise the impacts and disruption to road users. This gave the benefit of avoiding costs for years of traffic management, speed limits, delays and safety hazards which would be seen in a traditional solution. The 12 000-tonne bridge 'box' structure was built on land next to the M42, the box was constructed and, using a 'box jacking' technique, was sled in place in two different possessions/road closures at Christmas 2021 and 2022, minimising risks, disruption and ensuring safety for road users and so on.

Case study – SCAPE coastal defence projects

The obtained data from the SCAPE framework reviewed in Chapter 5 showed that the majority of projects were civil engineering and infrastructure projects. A notable pattern of projects was sea defence projects such as Broughty Ferry flood prevention,[1] Central Rhyl coastal defences,[2] Lowestoft flood risk management scheme[3] and Central Prestatyn coastal defences.[4]

Coastal defence projects by their very nature require action to be undertaken quickly to prevent flooding and damage. Therefore, ECI has been used to mobilise Contractors quickly and to develop innovation early in the project to overcome technical engineering challenges and selection of materials.

Balfour Beatty commented on this use of ECI on coastal defence projects as follows.

> Frameworks such as the SCAPE Civil Engineering framework and the Coastal Partnership East also give customers unique access to our team of unrivalled experts – they know that we can deliver and are up to the job.

> Critically, these frameworks also enable early contractor involvement which helps us to implement innovative technologies and solutions from the outset, to standardise specifications and undertake schemes simultaneously whilst also allowing for shared resources and ensuring effective reuse of materials (Mumford, 2023).

Notes

1. Broughty Ferry flood prevention undertaken by McLaughlin & Harvey under the SCAPE Major Works UK framework (McLaughlin & Harvey, 2024).
2. Central Rhyl coastal defences undertaken by Balfour Beatty under the SCAPE National Civil Engineering and Infrastructure framework.
3. Lowestoft flood risk management scheme – tidal flood wall – phase 3 and 4 undertaken by Balfour Beatty under the SCAPE National Civil Engineering and Infrastructure framework.
4. Central Prestatyn coastal defences undertaken by Balfour Beatty under the SCAPE National Civil Engineering and Infrastructure framework.

REFERENCES

Banwell CH (1964) *The Banwell Report: The Placing and Management of Contracts for Building and Civil Engineering Works*. Her Majesty's Stationery Office, London, UK.

Chandler D (2020) *Study Identifies Reasons for Soaring Nuclear Plant Cost Overruns in the US*. Massachusetts Institute of Technology, Cambridge, MA, USA.

Cheung D (2019) *Involving Contractors Early Using NEC*. https://www.neccontract.com/news/involving-contractors-early-using-nec (accessed 24/12/2023).

Designing Buildings (2020) *Optimised Contractor Involvement*. https://www.designingbuildings.co.uk/wiki/Optimised_contractor_involvement (accessed 17/07/2024).

EDF Energy (2014) *Experiences with the construction of EPR at Flamanville and Taishan OECD/Nuclear Energy Agency Workshop: Project and Logistics Management in Nuclear New Build*. https://www.oecd-nea.org/upload/docs/application/pdf/2020-07/wpne_workshop_3._1_experiences_with_the_construction_of_epr.pdf (accessed 12/07/2024).

FIDIC (Fédération Internationale des Ingénieurs-Conseils / International Federation of Consulting Engineers) (1999) *Yellow Book* (Plant and Design Build) FIDIC, Geneva, Switzerland.

Heier Sødal A, Lædre O, Svalestuen F and Lohne J (2014) Early contractor involvement: advantages and disadvantages for the design team. *Proceedings of the 22nd Annual Conference of the International Group for Lean Construction, Oslo, Norway*, p. 526. https://www.researchgate.net/publication/267210780_Early_Contractor_Involvement_Advantages_and_Disadvantages_for_the_Design_Team (accessed 24/12/2023).

HS2 (High Speed Two Limited) (2023) *Marston Box Rail Bridge.* https://www.hs2.org.uk/building-hs2/viaducts-and-bridges/marston-box-rail-bridge/ (accessed 12/07/2024).

Kan D (2018) *NEC4 X22: More Contractor Incentive in Two-Stage Contracts?* https://constructionmanagement.co.uk/nec4-x22-more-contractor-incentive-two-stage-contr/ (accessed 24/12/2023).

McLaughlin & Harvey (2024) *Broughty Ferry Flood Protection Scheme, Dundee.* https://www.mclh.co.uk/project/broughty-ferry-flood-protection-scheme/ (accessed 12/07/2024).

Moore C (2023) Construction of Hinkley Point C nears first reactor milestone. *New Civil Engineer*, 12 September.

Mosey D (2009) *Early Contractor Involvement in Building Procurement: Contracts, Partnering and Project Management.* Wiley Blackwell, Hoboken, NJ, USA.

Mumford P (2023) Early contractor involvement in coastal defence schemes is key to finding the best solution. *New Civil Engineer*, 18 September.

NEC (2016) *NEC Plays a Key Role on Hinkley Point C Project.* https://www.neccontract.com/news/nec-plays-a-key-role-on-hinkley-point-c-project (accessed 12/07/2024).

NEC (2017) *NEC4: Engineering and Construction Contract.* NEC, London, UK.

NEC (2020) *Preparing an Early Contractor Involvement Contract.* NEC, London, UK.

NEC (2022) *NEC releases new ECI Practice Note.* https://www.neccontract.com/news/nec-releases-new-eci-practice-note (accessed 24/12/2023).

Purcell M (2022) *A White Paper: Exploring Contract Models.* https://www.criticalinput.com.au/a-white-paper-exploring-contract-models/ (accessed 24/12/2023).

RIBA (Royal Institute of British Architects) (2020) *Plan of Work 2020 Overview.* RIBA, London, UK.

Walker A (2022) *Megaprojects Under the Microscope at London Contracts Conference.* https://infra.global/megaprojects-under-the-microscope-at-contracts-conference/ (accessed 12/07/2024)

emerald
PUBLISHING

ice
Publishing

Michael Smith, Matthew Finn and Jon Broome
ISBN 978-1-83549-897-2
https://doi.org/10.1108/978-1-83549-894-120241008

Chapter 8
The *Project Manager's* perspective

Dr Jon Broome

Abstract

This chapter describes

- the tightrope: impartial contract administration against project managing on the Client's behalf
- being empowered
- key competencies of the Project Manager and the team
- starting the contract off on the right foot.

8.1.   Introduction

As per Chapter 6 on the *Client's* perspective, I have also not been an NEC Project Manager. However, I have mentored, coached and provided support to many and their teams, both in a proactive manner to help set projects up for success and in a reactive manner when things are not going so well. So, this chapter is a collection of the four most important 'high-level' lessons learned, not from personal experience as an NEC Project Manager, but from assisting many.

8.2.   The *Project Manager's* tightrope: impartial contract administration against project managing on the *Client's* behalf

Under all editions of the NEC Engineering and Construction Contract, the *Project Manager* has a tricky role to play in that, in some circumstances, he or she is meant to be an 'impartial administrator' of the contract, while in other circumstances, they represent the *Client's* best interests. Where do these two roles come from?

The first thing to say is that no such roles are stated in the contract, rather individual responsibilities – a level down in detail from high-level roles – are stated in the contract in the individual clauses, so the *Project Manager* needs to know what mindset to adopt for each specific clause. So, let's explore what these roles mean in practice and when to have the particular mindset.

The stated role of representing the *Client's* best interests comes from the Guidance Notes, which were published with the contract up to the fourth edition when they became User Guides. In these Guidance Notes, it stated this was the role of the *Project Manager*. My rule of thumb – which is largely but not absolutely true – for when the *Project Manager* can make decisions and act in the *Client's* best interests is whenever there is a 'may' in the sentence, for example 'The *Project Manager* may' … 'propose … an acceleration' (clause 36.1)/'propose that … the Scope should be changed so that a Defect does not have to be corrected' (clause 45.1)/'instruct … alternative quotations' (clause 62.1) and so on (NEC, 2017).

In which case, where does the requirement to be 'an impartial administrator' come from? If you go back far enough, it comes from how contracts are ultimately interpreted in a court of law, whereby it would be a reasonable reading of the words by 'an objective third party' – that is, a judge. Therefore, to avoid a dispute, the *Project Manager* needs to interpret and act this way too. The specific case law on NEC which confirmed this was the first case law on the NEC in May 2005 in which Judge Jackson stated with regard to the specific issue over certification of payment

> I am unable to find anything which militates against the existence of a duty upon the project manager to act impartially in matters of assessment [of compensation events] and certification [of payments and more generally] when the project manager comes to exercise his discretion in those residual areas [of subjectivity in the NEC] … it would be a most unusual basis for any building contract to postulate that every doubt shall be resolved in favour of the employer and every discretion shall be exercised against the contractor (*Costain* v. *Bechtel*, 2005).

What this means in practice is that the *Project Manager* interprets the contractual reasons for nonacceptance as a reasonable person, experienced in that field, would do, based on the information available to him or her which includes but (as lawyers would say) 'is not limited to' that supplied by the *Contractor*.

To give an example to illustrate this: under a cost-based contract, the *Contractor* is missing a delivery note for, say, reinforcing bar. It would not be reasonable to refuse payment if the *Project Manager* – or more likely the quantity surveyor (QS) – could step outside and onto site to see a large pile of rebar clearly marked up as relating to that which the *Contractor* is claiming for payment. Other supporting information, such as photographs, are also relevant.

My rule of thumb for when the *Project Manager* should act as an impartial administrator is whenever he or she has to accept or not accept something, with the criteria for not accepting something stated in the clause. This applies to *Contractor's* design; the *Contractor's* programmes submitted for acceptance; whether an event is or is not a compensation event;

the evaluation of a quotation or making his or her own assessment and other circumstances. The exception to my rule of thumb is for certification of payment but, given that the case law was specifically on this point, I think we can take that as read.

To illustrate the difference between administrating the contract and project managing it, let me give you a simple example. Let us say it is on a school project, which has to be open by the start of term. A compensation event occurs which would push the contract Completion Date beyond this date, with the planned Completion on or around this date.

▨ Having had discussions with the *Contractor* about different ways of dealing with the event (as required under NEC, 2017, clause 62.1), the *Project Manager* instructs a number of quotations based on these different ways of dealing with it. Let us say one is for least time (crashing the programme with people working on top of each other); one for least construction cost which, when worked through, would mean the school probably opens late; and one somewhere in between, so if there is no other significant delay, the school will just be ready for term time. In instructing alternative quotations which give the *Client* different options, he or she is acting in the *Client's* best interests.

▨ In assessing each of the *Contractor*'s quotations, the *Project Manager* has to be an impartial administrator when evaluating them – that is, what would an impartial person of reasonable experience decide, based on the information presented in the quotation, other information readily available to them (such as the programme, Defined Cost information they already know about and looking around the site) and against the contractual criteria in clause 63?

▨ Having done this, the *Project Manager*, now acting in the best interests of the *Client*, can choose which of these quotations to instruct and implement.

Broadening this out, the impartial administrator role would almost certainly apply to the *Supervisor* when evaluating the quality of work that the *Contractor* has produced against the requirements stated in the Scope.

Further, it is only on the smallest contracts that you have a single person who does all the work of the *Project Manager* (and sometimes *Supervisor*), so the whole supporting team need to know when they are to act as impartial administrator and when in the *Client's* best interests.

And lastly, if the ECI stage one is being done under a professional services contract, whether acting as the *Client's* delegate (under the NEC4 Professional Service Short Contract), *Service Manager* (under the NEC4 Professional Service Contract) or other arrangement, the relevant individual needs to have this in mind too.

Under any of the above, if the relevant people are consistently making biased decisions, then

▨ not only are they putting the *Client* in legal peril should a dispute get to adjudication or beyond – that is, arbitration or the courts, but also

▦ they will be significantly undermining the stated 'spirit of mutual trust and cooperation' (NEC, 2017, clause 10.2) of the contract, which is potentially one of the reasons for using NEC, as well as choosing the ECI route.

The individuals need to be aware of this and what it means in practice. However, my experience is that, on some contracts, *Clients* do not make it easy for the contract team to act in this manner, which leads on to the next section.

8.3.   Good governance: having an empowered and competent contract team

It seems to me that an ever more common issue is the disempowering of the 'acting' *Project Manager* and team to act professionally and to make impartial decisions, all in the name of good governance. And the larger the project, generally the more scrutiny it comes under, so the greater the tendency for this to happen. And seeing as ECI arrangements are generally used for larger, more complex projects with more stakeholders, it is therefore particularly relevant to ECI contracts.

Let me give you two examples, although I could give a lot more.

▦ On a very high-profile project at a UNESCO World Heritage Site, the *Project Manager* and team were given very limited authority to instruct additional or changed works (of which there was a lot as it was a renovation project with limited ability to access and do good surveys beforehand). Further, no-one in the project team had the organisational authority to agree any compensation event, howsoever small. In fact, for a compensation event to be implemented, you had to go up three levels from the *Project Manager* to the Capital Investment Director (CID) for the whole estate. As a result, lots of side deals were being done for the *works* to progress, which were then wrapped up to be signed off by the CID who had neither the time nor knowledge to know if it was correct or not. Quite apart from not complying with the impartial administrator requirement – see previous section – is this good governance?

▦ On one current multibillion contract, which was procured using ECI, those in the *Project Manager's* commercial team who are involved in the day-to-day administration do not have the authority to implement any compensation event. Instead, having informally agreed it with the *Contractor's* personnel, it is then submitted formally under the contract by the *Contractor* and has to be referred upwards for review by a panel who almost always want some further detail or disagree with it in some way. The net result is that the already extended timescales for the *Project Manager's* responses are rarely being met, neither Party really knows where they stand commercially and an ever-increasing bow wave of unimplemented compensation events will almost inevitably lead to, at best, a negotiated settlement and, at worst, a full-blown dispute. Further, whether as a negotiating or legal position, it will be the *Contractor* who will be able to demonstrate that it was the one substantially complying with the contract. So, the governance is actively undermining the effective management and administration of the contract to both parties' detriment, but especially the *Client's*. Again, is this good governance?

A side effect of this is that much more professional time is needed to administrate the contract by both parties, which ultimately costs the *Client* money. So, before I proceed, just imagine if this was reversed where, before any programme, notification or quotation was submitted by the *Contractor*, it had to be reviewed by a committee? Not only would the *Contractor* be putting itself in legal and commercial peril but *Clients* – especially if paying on an open-book basis – would be up in arms about the cost of this bureaucracy. So why, in the name of 'good' governance, is it OK for *Clients* to have this bureaucracy?

So, what should *Clients* do instead? As per my views on open-book accounting – see Section 6.10 of Chapter 6 – it's in the set-up, so here are my pointers.

- Employ cooperative, competent people who can do their job in accordance with the contract, namely
  - a programmer who is knowledgeable about projects of this sort and knows and understands the NEC programming requirements. Ideally, without undermining it being the *Contractor's* programme for which it is responsible, they can constructively contribute to the betterment of the programme, not just evaluate it for acceptance and delays caused by compensation events
  - design personnel who, without undermining it being the *Contractor's* design for which it is responsible, can constructively contribute to design, whether in stage one or two, as well as accept or not accept the relevant design in accordance with the contract
  - a quantity surveyor who can build up cost from first principles and understands Defined Cost, so can both certify amounts due and assess compensation events correctly, whether evaluating *Contractor's* quotations or, hopefully rarely, making a *Project Manager's* assessment and, in an ECI environment, during stage one, co-operate with the *Contractor* in the development of the stage two target Prices or Budget
  - a good project manager who has, at a minimum, appreciation of the above as well as other aspects of what makes a good project manager[1] and is also an accredited NEC Project Manager: the two things are not necessarily the same thing or even over-lapping! However, they are mutually reinforcing.
- Do a RACI[2] diagram and wherever possible make the Accountable person (who contractually accepts the communication) the same as the Responsible person (who does the work) and delegate in accordance with the contract – that is, they are an 'RA' person. This is to avoid the situation identified above, where communications have to be referred upwards for review and acceptance.

- As a *Client*, make it clear to these people that they are largely impartial administrators of the contract as per *Costain* v. *Bechtel* (2005), so a programme has to be acceptable, not perfect; the designer cannot insist on gold-plated designs which pander to their own preferences if it's not in the Scope; and a QS is not doing a good job if they are consistently forcing the *Contractor* down as far as possible in the assessment of a compensation event or pushing the boundaries of what is a Disallowed Cost. Their job is to make a fair and reasonable assessment in accordance with the contract. And no decision, which includes requiring unnecessary additional information as a delaying tactic, is a poor decision.

- Set up information flows for those who need to be Consulted and Informed. Ideally, RA people are proactively informed as opposed to having to go out and consult. What do I mean by this? For example, rather than a QS having to seek out the programmer to consult over whether a compensation event is on the critical path and therefore whether the *Contractor* is entitled to time-related costs, through regular briefings which inform them, the QS already knows.

- As a *Project Manager* and *Client*, have the courage to trust your people in the day-to-day, month-to-month management and administration of the contract. But monitor the contract both through 'hard' contractual compliance statistics and data and 'soft' behavioural ones. For the former, this might include the number of unresolved compensation events, the value of unresolved compensation events and time to accept, all of which should be readily available when using a cloud-based contract administration system.[3] For the latter, it might be through bimonthly surveys on how people feel the contract is operating. Together with talking to people, trends can be identified and problem areas recognised. On large contracts, this might be by geographical area. It could also take the form of periodic audits of payment process – again see Section 6.10 of Chapter 6 – as well as prompt and correct entries on the cloud-based communications system, level of detail in the programme acceptance, compensation event quotations and so on.

- Based on the above, take supportive and targeted action. For example, if compensation event quotations are not being agreed in one geographical area, find out why, including getting the *Contractor's* perspective. It could be because the *Contractor's* QS is consistently over-egging the quotations. Alternatively, it could be because the *Project Manager's* QS is consistently not accepting them for spurious reasons or, quite commonly, wants every detail nailed down for a forecast which will never be the case. Where possible, address this at a systems level, rather than a case-by-case level, so it's not personal. So, for example, if this is because expectations about the amount of detail required vary, sit down and agree it, rather than having the same arguments time and time again.

Doing the above will, in my view, give the *Client* far better outcomes and knowledge of where the contract is heading than disempowering their delivery team in the name of good governance.

The Luas Line extension was a contract to extend one of the tramways in Dublin in the mid-teens of this century. The first construction contract was to move the underground services in the way of the line either out of the way or deeper down so that the main construction contract for the line, including the foundations, was de-risked. It was one of the first, if not the first, NEC contracts let in Ireland and was let as an Option C contract. While reporting to the Project Manager, *I was effectively a 'coach' to the joint team doing joint trainings, audits for compliance and good practice, both by way of having remote access to the contract administration system and through periodic site visits.*

The Project Manager *noted that relationships between the two commercial teams was starting to become strained, largely because of lack of agreement of what was or was not a Defined or Disallowed Cost. The* Project Manager *expressed no view over who was in the right or wrong, but asked me to 'bang heads together' as he did not want the issue to persist and poison other areas of the contract where there was a good deal of cooperation and collaboration. So, I flew over and spent the morning and the early part of the afternoon with the two QSs.*

What emerged was that the Project Manager's *QS had put an industry standard document about cost control in the Works Information (as it was an NEC3 contract), which was a bit unspecific about what evidence and registers the Contractor had to show. However, he had certain ideas about what this meant in practice which did not correspond with the* Contractor's *QS's ideas. Further, while the raw data could be downloaded from the* Contractor's *cost management system, it was not the format the* Project Manager's *QS wanted and would require significant on-going time commitment to rework it.*

We basically spent a morning working through and understanding each other's positions, making compromises and putting in place actions to address genuine shortfalls and fixes so that the Project Manager's *QS could see the relevant information.*

And that was that: three-quarters of a day to do a targeted intervention which saved both parties numerous hours and days of argument, prevented the bad atmosphere from spreading to the rest of the contract and enabled the final account – along with a few thousand compensation events – to be settled within three months of Completion. From the Client's *and* Project Manager's *perspective, a good investment.*

8.4.   Selection criteria and incorporating *Contractor* promises into the contract

In the section above, I briefly touched on the sort of people I would want managing and administrating a contract. By and large, I would want similar people from the *Contractor:* competent, professional people who can co-operate in accordance with the contract.

In all NEC4 long forms – by which I mean those without 'short' in the title – there are entries in Contract Data part two for the tendering contractors, consultants or suppliers to enter the name, job title, qualifications and experience of their proposed '*key persons*'. It is quite common, almost usual, for *Clients* to specify what jobs they want to have as *key persons* and these people are evaluated as part of the quality submission, so profiles or curriculum vitae (CV) are also attached by the tenderer. Indeed, on an ECI job, this might well be given quite a high weighting in the selection of the winning Contractor. And on signing the contract, it becomes a contractual obligation to supply these *key persons*, although they can be replaced by people with 'as good' 'experience and qualifications' (NEC, 2017, clause 24.1). Later on in Contract Data part two, under X22: Early *Contractor* Involvement, are additional entries for 'The Stage One *key persons* are …', again with entries for names, job title, responsibilities, qualifications and experience.

The point I wish to make from experience is that often these *key persons* do not seem to have experience of doing the ECI-type things the *Client* wants to be done, in terms of gaining consents, taking the scheme through planning inquiries, stakeholder management, delivering social value and so on. And why should they, as they are contractor people! This results in frustration for the *Project Manager* and team as, rather than assisting the *Contractor* in doing the roles assigned to them under the contract, they end up leading or even doing them. Indeed, on one contract, we went so far as to take the work out of the Scope and hence reduced the target Prices, such was the extent of failure. Getting to this point creates a lot of frustration and tension within the project team, which undermines one of the major reasons for choosing the ECI approach.

So, how does it get to this point and what can the *Client* and *Project Manager* do about it?
- Firstly, at prequalification stage, the tendering contractors who then go forward to bid, all give shining examples of where it, as an organisation, has done these or similar roles before. In reality, these may not have been such beacons of best practice, and they create an expectation that the individual people supplied will be competent, which may not be the case. So, the first thing to do is to check out the reality of the examples given in the prequalification documents.
- Secondly, when specifying the *key person* job roles in the Contract Data, the *Client* only specifies the senior management roles and not the specific ECI roles – such as gaining consents, taking the scheme through planning inquiries, stakeholder management, delivering social value and so on – that they want to be performed. Consequently, there is no obligation to provide these suitably qualified and experienced people. Rather, someone will (hopefully) learn on the job!
- Thirdly, and closely related to the above point, the entry for 'responsibilities' is rather vague: I like to make specific reference to the relevant sections in the Scope. So, if there are roles which are unusual for a Contractor, which there often will be in an ECI arrangement, specify them (the point above) and specify them with precision.
- And fourthly and last of all, the Scope is somewhat vague about what the *Contractor* has to do. Often this is because the *Client* has specified at a high level what it wants the

Contractor to do and then asked a question, as part of the quality submission, about the specifics and how it will fulfil these high-level obligations. The tendering Contractors' responses read really well but, when analysed for tangible specifics which they can be held to account for, they are actually rather vague.[4]

The last point brings me onto the general, but very important point, I wish to make: as well as being scored, whatever promises the tendering Contractors (and ultimately winning one) make have to be capable of being incorporated into the contract. Otherwise, they are not worth anything legally, so why bother asking and scoring them!

Some context for the above paragraph: on complex projects – such as on live assets with multiple stakeholders where ECI is used – often the 'how' it is delivered is just as important as, and sometimes more important than, 'what' is ultimately delivered. Therefore, promises made at tender, which are fundamental to who the winning Contractor is, need to be incorporated into the contract and form part of the winning *Contractor's* obligations. This needs to be made clear at tender, in the Invitation to Tender documents, and the contract needs to be structured to do this.

Unfortunately, the unamended words of the NEC contract are, in my view, not quite right here. The relevant entry in Contract Data part two is 'The Scope provided by the *Contractor* for its design are …' However, promises about methodology are not design. You could say that these promises about methodology should be in the *Contractor's* programming submission and therefore reflected in the next entry in the Contract Data part two whereby their tender programme is referenced and incorporated into the contract. However, the *Contractor* can change this once the contract is entered into, with one of the reasons for not accepting the revised programme being 'it does not comply with the Scope'. But if the *Contractor's* tender promises are not part of the Scope, then this is not a valid reason for not accepting the programme and the *Contractor* is free to satisfy the *Client's* high-level (and therefore vague) requirements any way it decides.

Fortunately, the solution is rather easy. As an amendment to the contract, wherever the words 'Scope provided by the *Contractor* for its design' is used in the contract, replace this with the words 'Scope provided by the *Contractor*'. This then covers both design and methodology. And in Contract Data part two, I typically insert another entry which is 'The Scope provided by the *Contractor* for its methodology is in …' and, just to reiterate, make it clear that promises made at tender may well be required to be incorporated here.

8.5. Starting the contract off on the right foot

At the time of writing, 'collaboration' is the slightly declining in-fashion buzz word that will apparently lead to a successful contract. If collaboration is so important, then surely it makes sense to manage and plan in collaboration, rather than just hope or expect to it occur?

Having said this, collaboration, if viewed as behaviours only is, in my opinion, massively overrated and I can give numerous examples of project teams collaborating but little or nothing getting done. Worse still is when hard issues are brushed under the carpet as to raise them would be uncollaborative so, because they are not proactively addressed, end up having a bigger impact and cause the parties to fall out over who is to blame. Consequently, relying solely on a behavioural psychologist or equivalent profession to do training, workshops and mentoring will have limited affect.

Instead, if collaboration is seen as part of good project management in its widest sense, then to set the project and contract up for success, we want to set the project and contract up in accordance with good project and contract management principles (which, to be clear, includes collaborative behaviours). And as an observation, it is far easier to establish a culture of good project management at the outset and then embed it in the team, rather than change a failing culture.[5]

Rather than give you theory, let us look at an example taken from one of my regular clients, Connect Plus, from which I gave a number of examples for the different ECI types in Chapter 2.

As previously stated in Section 2.5, Connect Plus operate, maintain and improve the M25 London Orbital under a 30-year DBFO (design, build, finance and operate) contract on behalf of National Highways. This was awarded in May 2009. The 'operate and maintain' part of the contract, as well as the big improvements (measured in £100s of millions) in the early years of the contract, were and are paid for on a fixed price monthly payment basis regardless of what it costs Connect Plus or when it incurs the costs. However, subsequent improvements up to c. £20m are funded by National Highways with the money coming through Connect Plus who, with their consultants, are responsible and accountable for their successful delivery.

The previous COFA 2 arrangements for improvements, running from 2017 to 2023, contained all the contractual provisions for ECI, yet it was recognised that it was only in the later years of the 6-year framework duration that, despite having a highly collaborative culture established with the help of a psychologist, value from ECI had not been realised. Outcomes started to improve towards the end when a more disciplined approach to contract delivery was enacted and the aim for the succeeding COFA 3 framework was to embed and further improve these. This included a detailed RACI diagram[6] at framework level for which party did what (and, if the Contractor, whether it was paid for directly or indirectly as part of their overhead by way of the *fee percentage*).

As described in the example in Section 2.5 for ECI type 1, each framework contractor had an annual NEC4 Professional Service Short Contract from which time charge orders, as compensation events, could be instructed for very early contractor involvement. This included attendance at a precontract meeting or workshop, together with the funder, Client, consultant and potential supply chain partners, to finalise the ECI stage one Scope. The workshop's purpose was two-fold and, to quote from the contract documents

> "The purpose of the start-up workshop is that the principal stakeholders/participants
> * have a similar knowledge and understanding about the proposed Call-Off Package. This includes the benefits delivered from a successful project, success criteria for the delivery phase, the main challenges/issues, threats, opportunities and assumptions and surfacing the less obvious ones
> * agree and commit to the key activities and dates necessary to agree the quotation for stage two to proceed in a timely manner."

For the first bullet point, it then gives a structured list – which matches the points listed, but in more detail – of what is covered to 'orientate'[7] participants to the project. The second bullet point effectively gives the deliverables from ECI stage one which are necessary to proceed to stage two. The document also gives detailed information on the contents of *Contractor's* submission in terms of programme, financial submission and Early Warning Register. And lastly, it describes in some detail how and when the parties are to interact, both on an informal ad-hoc basis and formal planned basis over the duration of stage one.

All this is a far cry from the high-level expression of collaboration with no detail on the bones which is what I see in the literature and other articles. In fact, I do not think you can even say there are any bones or skeleton on most of what I read about collaboration! Instead, my experience is that it is the detailed application of good project and contract management principles that, unsurprisingly, delivers successful projects and contracts.

To conclude
* am I saying you should do exactly as Connect Plus have done and are doing? No. For a start, your circumstances might be different, which may include existing processes and practices which work for you.
* am I saying you should think about the specifics of collaboration and include the results of your thinking, which should be tangible deliverables and actions, somewhere in the contract documents? Most definitely 'Yes'!

8.6. Conclusions

This chapter has covered
* the importance of the *Project Manager* and his or her team being aware of when they are to act in the *Client's* best interests and when they are to be 'impartial administrators' of the contract

▨ that the *Client* needs to be aware of this and empower the team to act in this way rather than require that decisions should be referred upwards. Among other things, this slows decisions, increases bureaucracy for both parties and puts the *Client* in legal and commercial peril – that is, undermines the likely success of the contract. Instead, the *Client* and the *Project Manager* should employ competent, cooperative professional people who understand the contract. They should then create the structures and information flows for them to operate effectively and monitor performance in order to take targeted, supportive action.

▨ that it must be made clear to the tendering Contractors what they have to do in stage one, that promises made at tender will be incorporated into the contract and that this includes *key people* who have experience of doing ECI type work which may be untypical for a typical Contractor. The *Client* needs to ensure that this work is stated in sufficient detail in the contract and that *key persons'* jobs and responsibilities are cross-referenced with it. Otherwise, the *Project Manager's* team will not be able to enforce these (nonexistent) obligations.

▨ that just being collaborative alone is not sufficient. Good project and contract management and administration practices and processes need to be put in place early on in the contract to get the best from the use of ECI. These need to be expressed in the contract and some sort of project/contract start-up event needs to be held to get everybody on the same page about what the project is about and what needs to be achieved for stage one to be successful.

The common theme among all the above is that, at a minimum, the expectation is created precontract, enshrined in the contract and action is taken early on in stage one. Failure to do this puts the *Project Manager's* team on the back foot, hindering the effective operation and hence delivery of the contract. On the other hand, proactively addressing these issues does the opposite.

Notes

1. For me, the gold standard qualification which requires knowledge, understanding, experience and achievement is the UK's Association for Project Management's (APM) Chartered Project Professional (ChPP) qualification. It is fast becoming the requirement to manage major projects in the UK and may well expand out to other jurisdictions. But note that I am an ex-trustee and deputy chair of the APM, so may be biased!

2. RACI stands for Responsible, Accountable, Consulted and Informed. The Responsible person is the person who actually does the work, whereas the Accountable person is the person who can be held to account. Without any delegation, the Accountable person under the NEC4 Engineering and Construction Contract is the Project Manager.

3. Ideally, this NEC contract administration system should be able to interface seamlessly with your project controls system, giving real-time bottom-up knowledge and insight into

where the contract is heading and, by combining it with other contracts, where the project is heading. Unfortunately, despite there being a multitude of very good NEC contract management systems out there, to my knowledge at the time of writing, only two have the potential to do this.

4. I was once part of the team evaluating Contractor responses for an ECI contract. Even for the winning Contractor, while the layout, presentation and how the response read was great, when I actually went through highlighting the text that was sufficiently precise to be contractually enforceable, less than 10% of it ended up being highlighted. This was despite it being emphasised in the tender meetings that this was important and something like 'using language which is specific enough to be contractually enforceable' being part of the scoring criteria.

5. If you want to see myself – and Lesley Hunt of Jacobs – really have a go at 'collaboration' and emphasise the importance of good project and contract management (and what it is) then there is a video which can be found here: https://vimeo.com/875612044?share=copy (Hunt and Broome, 2023).

6. https://www.jonbroome.com/resources/videos/nec3/should-we-co-operate-or-collaborate-or-just-no.

7. I use the term 'orientate' as an alternative to the 'forming, storming, norming and performing' model of team development which was based on observation of un-managed team development. Instead, the storming part is replaced with a managed 'orientating' and 'creating' phase, i.e. you orientate people to what the project is about, so participants are on the same page about where the project has come from, and then jointly and consciously create the objectives and how you are going to work together to deliver them. These then become the norm for the project team to perform. For a more detailed article on this, go to my website: https://www.jonbroome.com/getattachment/c683ec16-55c2-4e45-a5da-9a8c8b8d48c0/How-to-stop-your-team-'Storming'-and-get-them-'Per.aspx.

REFERENCES

Broome J (2008) *How to Stop Your Team Storming and Get Them Performing.* https://www.jonbroome.com/getattachment/c683ec16-55c2-4e45-a5da-9a8c8b8d48c0/How-to-stop-your-team-'Storming'-and-get-them-'Per.aspx (accessed 23/05/2024).

Costain Ltd v. *Bechtel Ltd* (2005) EWHC 1018.

Hunt L and Broome J (2023) *Cooperate, Collaborate or Not Bother.* NEC People Conference. https://vimeo.com/875612044?share=copy (accessed 12/07/2024).

NEC (New Engineering Contract) (2017) *NEC4: Engineering and Construction Contract.* NEC, London, UK.

emerald PUBLISHING ice

Michael Smith, Matthew Finn and Jon Broome
ISBN 978-1-83549-897-2
https://doi.org/10.1108/978-1-83549-894-120241009

Chapter 9
Conclusion and key points

Michael Smith, Matthew Finn, Dr Jon Broome and Catherine Maddox

In this book we have sought to evidence how a properly constructed and administered two-stage procurement process can deliver real benefits for both the Client and the Contractor on a project where the design solution and pricing proposal, which deliver best value for money, are uncertain at the outset. By way of a conclusion, here are our top tips to help you structure and administer such a process in a way which should give you the best chance of realising those benefits.

9.1. From a legal perspective

- Use a tried and trusted standard form contract, containing the requisite project management tools, as the base document for the conditions of contract: we suggest using the NEC4 Engineering and Construction Contract, incorporating the main pricing Option C (Target Contract with Activity Schedule) and the secondary Option X22 (Early Contractor Involvement) (NEC, 2017).
- Resource accordingly – the success of this labour-intensive procurement option will largely be dependent on the quantity and quality of the client's management resource.
- Consider contractor risk appetite and market conditions within the relevant industry sector: will a fixed price or a target price during stage two deliver best value for money?
- Safeguard against the risk of programme slippage and a gradual erosion of the Client's bargaining power during stage one by
  - providing clearly for the client's right to withdraw from the process without penalty at the end of stage one, to proceed with another contractor and to use the design information provided by the Contractor (on the basis that the Client has made payment for it)
  - requiring agreement on the stage two conditions of contract as a condition precedent to the stage one appointment
  - appointing a strong design consultancy to assist the Client during stage one, with a remit to
    - investigate, analyse and develop the Client's design requirements and the Contractor's proposals
    - manage a robust value engineering process throughout

- o maintaining competitive tension within the tender procedure by
  - ▪ evaluating change to the stage two target Prices using the competitive pricing information submitted with the Contractor's original bid submission
  - ▪ allowing the Contractor to share in any saving between the initial target Prices included within the Contractor's accepted tender and the sum of the revised target Prices agreed at the end of the stage one process and the amounts paid during stage one (albeit with payment of any such saving held over until completion of the works and the assessment of any pain share payable by the Contractor at that stage)
- ▦ If using Option X22 together with the main Option C target Prices pain/gain share arrangement, consider the content of the two incentive schemes – how will they best work together to promote consensus and avoid unnecessary costs escalation during both stages?
  - o Consider what elements of project cost can be influenced by the Contractor and should therefore be included within the Budget (by reference to which the Budget incentive payment is calculated) and the circumstances in which that Budget is then adjusted in order to preserve the incentive.
  - o As regards the target Prices pain/gain share arrangements during stage two, consider
    - ▪ those cost items which will fall within the definitions of 'Defined Costs' and 'Disallowed Costs' and the extent to which the Client should have control and/or oversight of the Contractor's supply chain arrangements
    - ▪ whether, for the purposes of calculating the Contractor's pain or gain share
      - ☐ the estimate of the Contractor's total Fee entitlement should be included within the target Prices set out within the Activity Schedule (given that the Contractor's actual Fee entitlement is included within the Price for Work Done to Date)
      - ☐ any account should be taken of any key performance indicators (KPI) or other deductions that are made from the Price for Work Done to Date
    - ▪ whether, once the Price for Work Done to Date has exceeded the target Prices (or there is no reasonable prospect of the Price for Work Done to Date being less than the target Prices), the Project Manager should be able to deduct from all future payments amounts equal to its estimate of the Contractor's ultimate pain share
    - ▪ whether any compensation events which reduce the Contractor's Defined Costs should also reduce the target Prices
    - ▪ the extent to which recoveries made by either party under project insurances taken out by the Client should be taken into account for the purposes of the pain/gain share calculation
    - ▪ the percentages at which and the bands within which any pain and gain is shared between the parties
    - ▪ any circumstances in which the Fee should be no longer payable on the Contractor's Defined Costs
    - ▪ the extent to which any Contractor exposure to pain share should count towards any limit on its aggregate liability under the contract.

▦ Consider how best to manage package interface risk and avoid consequent costs escalation.
   o Does the Client (or its construction manager), the Contractor and its supply chain have a track record in the management of such risk and are there adequate sanctions and/or incentives to ensure that the Contractor complies with its project management-related obligations?
   o Are there clear and consistent contractual provisions relating to the allocation and management of construction risk across all the main packages (including the use of information release schedules, key dates regimes and operational interface protocols)?
   o Is there a detailed strategy for the management of particular project delay risks (including available insurance cover)?
   o Has there been enhanced technical adviser due diligence and modelling on the likelihood of risk occurrence and do the programme and budget contain appropriate consequent contingencies?

9.2.   From a tender competition perspective

▦ Identify whether or not public procurement law applies to the contract in question.
   o Is the awarding body a contracting authority?
   o Is the contract value above the threshold to which public procurement law applies?
▦ If a public procurement procedure needs to be run, consider the appropriate choice of public procurement procedure, including
   o whether there is a need for the contracting authority to have more involved discussions and/or negotiations with bidders about the contract requirements and their proposed solutions
   o whether there is a need for dialogue or negotiation: either the competitive dialogue or competitive procedure with negotiation (Public Contracts Regulations 2015, HMG, 2015) should be used
   o Alternatively, if the Procurement Act 2023 (HMG, 2023) has come into force, the competitive flexible procedure should be utilised, with the addition of some dialogue or negotiation stages.
▦ Determine the selection criteria and award criteria for the public contract. In particular, consider the following issues when choosing award criteria.
   o The criteria must be linked to the subject matter of the contract.
   o The authority must choose criteria that identify the most economically advantageous tender or, if the Procurement Act 2023 has come into force, the most advantageous tender.
   o The award criteria used must almost always include an assessment of price or cost.
   o Has the entire scope of the two-stage contract in fact been procured compliantly?
▦ The contracting authority must consider whether the pricing evaluation is sufficient.
   o The most compliant public procurement approach for a two-stage contract is to request pricing for both stages of the contract, both the initial stage and the subsequent stage.

o   If an insufficient range of pricing metrics is taken into account, it is difficult for the contracting authority to reach any objective determination as to which tender was the most economically advantageous. Therefore, an exercise will be required to ensure that sufficient and comprehensive pricing criteria have been included, some of which must also apply to the second stage of the contract.

o   One common public procurement approach is to ask Contractors to price sample schemes and then apply the principles of this pricing to the pricing for the contract.

o   Other approaches are to include provisions in the contract to ensure that payment for the second stage of work does not alter the economic balance of the contract; for example, that it does not alter the risk allocation or increase the Contractor's profit margin.

o   Whether or not the price evaluation approach chosen will be legally sufficient very much depends on the contract in question and the extent and range of pricing metrics evaluated during the public procurement stage.

o   Although the exact pricing assessment metrics chosen will be specific to the contract in question, it is clear that a broad range of pricing metrics, covering pricing for both stages one and two, will be required in order for contracting authorities to be able to argue that pricing for the full scope of the two-stage contract has been adequately tested and competed.

9.3.   From a commercial perspective

▨   Understand when and where ECI is likely to work or not (see Section 2.2 in Chapter 2).

▨   Consider the general advantages and disadvantages of generic ECI arrangements (see Figure 2.4 in Chapter 2). In summary, one could consider using an ECI approach when

o   the project has sufficient complexity, risk and/or monetary value to justify the expense of paying for additional professional time up-front, as a value-adding investment – that is, issues will be resolved, downside risks reduced or eliminated and opportunities taken advantage of

o   there is sufficient time in the project timescale before stage two – when the big money is being spent – to address the issues, risk and opportunities during stage one

o   despite the investment of time in stage one, issues, risks and opportunities (which both parties can contribute to) will still be encountered in stage two to make it worthwhile using a pain/gain arrangement for stage two to stimulate collaboration. If not, one could still take an ECI approach, but with a lump sum priced contract for stage two

o   the Client and the sector in which it operates have the maturity of skills, both organisationally and individually, and behaviours to work together in a value-adding way as opposed to reverting to type. This might not be overtly confrontational but could just be a man-marking 'jobsworth' manner.

▨   Consider which of the six basic types of ECI (summarised in Section 2.4 of Chapter 2) might be the most appropriate starting point for the procurement model (see Sections 2.5–2.10 in Chapter 2) but bear in mind that

o   all have their advantages and disadvantages, which need to be matched to the project circumstances in order to maximise the pros and minimise the cons o all six options are archetypes, but none are fixed in stone and hence can be adjusted, evolved and, in some cases, combined to suit a project's or a programme's particular circumstances.

- Consider whether one of the three basic forms of project alliance (see Section 2.11 in Chapter 2) is appropriate.

- Learn from the client insights referred to in Section 6.2 in Chapter 6: make sure that objectives are not expressed too vaguely or do not stand up to scrutiny when challenged, particularly around the desire for certainty. Be specific, as well as truthful and honest, about what your objectives for the project and hence contract are.

- Consider how best to balance value for money with price certainty and affordability (see Section 6.3 in Chapter 6) which inevitably will lead to a consideration of the relative risk and reward (see Section 6.4 in Chapter 6). While 'allocating risk to the party best able to manage', whether threat or opportunity, is certainly a principle worth considering, so too is the effect on that party's bottom line. Allocating risk to a party that will become insolvent if the risk occurs is unlikely to be good business.

- If a pain/gain share incentive regime is to be adopted, consider the respective share percentages and the circumstances in which the Budget or target Prices are adjusted (see Section 6.6 in Chapter 6). These are the two main factors that are likely to determine whether there is a market for the incentive and if so, how it will be priced. So, prior market consultation is to be encouraged (see Section 6.7 in Chapter 6).

- Given that the target Prices or Budget for stage two will be determined in the absence of a competitive process, clearly document the process and level of information that is required to arrive at such Prices or Budget (see Section 6.8 in Chapter 6). The most contentious area is likely to be the extent to which risk allowances are priced, so this needs to be thought through and prescribed in particular detail (see Section 6.9).

- Consider how best to administer open-book accounting on a cost-based contract, in both stages one and two. The cost of doing so falls largely on the Client who is paying for Client-side commercial staff and auditors as well as the Contractor-side commercial staff to present it. So, the Client's expectations should be clearly stated. Stage one should then give both parties sufficient time to understand processes, verification of cost of resources and how they are accounted for, before the big costs are incurred. This time should not be wasted.

9.4.   From the Project Manager's perspective

- When administering an ECI contract, the Project Manager should bear in mind the four 'high level' lessons learnt, which are explained in Chapter 6.

   o The Project Manager has a duty to both represent the Client's best interests and, when exercising a discretion under the contract, to act as an 'impartial administrator' of the contract.

   o Persuade the Client to adopt expedient and effective governance procedures which do not undermine the Project Manager's ability to manage the contract in an efficient manner. In particular

- employ cooperative, competent people who can do their job in accordance with the contract
- do a RACI diagram: Accountable people who draft and send communications required under the contract; Responsible people who can draft, but not send, these communications; who needs to be Consulted and Informed?
- set the Client expectation that, in the main, all participants are to act as impartial administrators of the contract
- set up information flows for those who need to consulted and informed, both through 'hard' data flows and 'face-to-face' briefings
- trust all participants, on a day-to-day, month-to-month basis, to do their job, but monitor their performance against both hard contractual and soft behavioural measures
- based on the above and a consideration of the Contractor's perspective, take supported and targeted action.

o Incorporate the promises made by the winning Contractor in its bid submission into the contract so they become contractual obligations (see Section 8.4 in Chapter 8). This not only applies to the people put forward, but also to the Contractor's answers to the quality questions.

o Provide the collaboration infrastructure. Ensure that the project and contract management infrastructure – organisation structure, responsibilities under a RACI diagram, meetings structure, processes and systems – are specified to a certain level of detail in the contract documents and are then implemented and refined during stage one. That way, when the big money starts to be spent, the team can be in performing rather than storming mode.

REFERENCES

HMG (Her Majesty's Government) (2015) Public Contracts Regulations 2015. The Stationery Office, London, UK, Statutory Instrument 2015 No. 102.

HMG (His Majesty's Government) (2023) Procurement Act 2023. The Stationery Office, London, UK.

NEC (New Engineering Contract) (2017) *NEC4: Engineering and Construction Contract*. NEC, London, UK.

Michael Smith, Matthew Finn and Jon Broome
ISBN 978-1-83549-897-2
https://doi.org/10.1108/978-1-83549-894-120241010
Emerald Publishing Limited: All rights reserved

Index